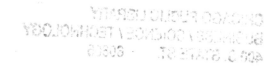

COPYRIGHT 1993 BY THOMSON PUBLICATIONS

Library of Congress Catalog No. 64-24795
Printed in the United States of America
ISBN - Number 0-913702-40-4

THIS BOOK FOR REFERENCE ONLY

READ THE LABEL CAREFULLY

TABLE OF CONTENTS

Page 15

Page 31

Page 67

Page 87

Page 159

Page 185

Page 197

Page 271

Page 297

INTRODUCTION

This book is designed as a helpful guide to all the available agricultural chemicals and their uses. Farmers, agronomists, pesticide salesmen, pest control operators, greenskeepers, conservationists, county agents, research workers, and many others should find this a handy reference for frequent use.

In this book the author has attempted to list and describe the most widely used agricultural chemicals marketed in the world today. Many of the compounds listed are not available commercially, as they are still in the experimental stages; however, by mentioning them, the life and usefulness of this book will be prolonged.

Since research is continuing on most of the chemicals available in the agricultural field, a check with your farm advisor, county agent, pesticide supplier, agricultural college, or other leading authority would be advisable before using the chemicals as recommended. The facts have been presented as they appear to the author, but this information is continually changing as knowledge is gained on the use of these materials.

HOW TO USE THIS MANUAL

The following indicate what is to be found under each heading for the individual chemicals listed. The chemicals are not listed alphabetically, but placed in groups of related compounds. All names, both chemical and common, are listed in the indexes appearing in the front of the book.

NAMES: An attempt has been made to list all the usual names by which a chemical is known. Since this manual is designed for use mainly by the non-technical individual, the common and trade names are used primarily. The most popular trade or common name is listed first, while the remaining are placed at random, not according to usage. The official common name, if the chemical has one, is italicized.

Immediately under the names appears the structural formula, if such a formula exists, followed by a written chemical formulas, for the sake of simplicity usually only one appears.

ORIGIN: The company which has done most of the development work on the compound is listed, followed by the year the compound was patented or put on the market. The chemical company mentioned may or may not have patented the compound, but it has done much of the basic research connected with developing it. On compounds developed outside the United States, the company which is licensed to market the chemical in this country is mentioned. On some of the older chemicals, only the principle basic producers are listed.

Since many companies may formulate the same chemical compound, readers should not be led to the believe that the company name is the only producer. The author is by no means trying to endorse certain companies' compounds

TOXICITY: The LD_{50} values of the different compounds are listed under this heading. Since they vary considerably with different tests conducted by different personnel, the lowest value found in the literature is given. The values are listed as the acute oral LD_{50} of the technical material, usually recorded in milligrams per kilogram of body weight (mg/kg). As most toxicology work is done with white albino rates, the value listed is that of the toxicity on them unless otherwise stated. Other information on toxicity may be mentioned.

FORMULATIONS: This category explains the forms in which a chemical is marketed. They are usually wettable powders (WP), emulsifiable concentrates (EC), oil solutions, granules, dusts, or aerosols. Since these formulations are constantly changing, the reader is advised to check into the different formulations available from his supplier in his particular locality.

PHYTOTOXICITY: Many pesticides, when applied to some plants, show detrimental side effects. The author has endeavored to list as many plants injured by a certain compound as it was possible to find. This changes in many cases due to the advent of new, less phytotoxic formulations. Weather conditions in a certain locality may also be a factor. Many of the chemicals are still in the experimental stage, so phytotoxicity data on them may not be complete.

USES: Since this manual was designed primarily for use in the United States, the plants and animals for use on which the different pesticides are registered by the Environmental Protection Agency and the United States Department of Agriculture are listed under this heading. This can serve as a guide for usage outside of the United States. An attempt was made to list all the registered uses at the time of publication, but many uses may have been deleted or added since then. Your farm advisor, county agent, or chemical supplied can also give you proper verification before applying to a questionable usage. A good rule to remember is: don't use a pesticide if the desired usage is not referred to on the chemical label unless the feasibility of such usage is confirmed by a reliable source. *Read the label carefully.*

Experimental uses, so designated, are mentioned for some of the newer compounds. The compounds may or may not be registered for such uses with the next few years.

RATES: The extremes of high and low dosages are stated in most cases on a per-acre basis and on a per-100-gallons-of-water basis. The average dosages used should normally fall in between these rates. Your supplier or county agent can give you much more accurate information on this for your specific locality or situation.

IMPORTANT DISEASES CONTROLLED OR PREVENTED: The heading is self-explanatory in that only the pests considered by the author to be of major importance are mentioned. In the interest of simplicity and space, not all the pests a certain chemical will control or prevent could be listed.

APPLICATION: This, at its best, is only a general guide to the application of these compounds. As locality, weather, rates, crops, etc., vary greatly, only a general recommendation can be given. Authorities in the reader's locality can give much more specific recommendations.

PRECAUTIONS: This is self-explanatory in that the possible hazards connected with the use or application of the specific compound are mentioned.

RELATED MIXTURES: Mixtures containing the compound previously mentioned are listed here, together with the company which produces them. Due to space restrictions, not all such mixtures could be mentioned, so only a few with registered trade names were selected. They are mentioned merely to let the reader know they are available and to avoid confusion of the different names.

RELATED COMPOUNDS: Pesticides which are closely related to the above compounds are listed. These, so listed, are used to a very limited extent, and do not warrant a full page. Only a brief description is given.

STATEMENT OF WARRANTY

The author and/or publisher are in no way responsible for the application, etc., of the chemicals mentioned. They make no warranties, expressed or implied, as to the accuracy or adequacy of any of the information presented in the writing.

NOTICE

Since new pesticides are constantly being introduced, a revision of this manual will be available when the number of such chemicals warrants it. As of this writing, there are a series of four manuals available: Agricultural Chemicals Book I, Insecticides and Acaricides; *Agricultural Chemicals Book II - Herbicides; Agricultural Chemicals Book III - Miscellaneous Chemicals, Fumigants, Growth Regulators, Repellents, and Rodenticides; and Agricultural Chemicals Book IV - Fungicides.* More information may be obtained by writing the publisher.

TRADEMARKS

The material in this book has been assembled from a multitude of labels, bulletins, instruction sheets, etc., published by the various companies for the public's information in the use of their products. For the sake of simplicity, reference to and use of registered trademarks has been eliminated. This should, by no means, indicate the absence of proprietary right on the use of such words. Also, by the omission of certain trade names, either unintentionally or from lack of space, the author should not be considered to be endorsing only the companies whose brand names are listed.

NAMES, COMPOUNDS & MIXTURES

B

C

H

I

S

PHENOXY COMPOUNDS, BENZOIC ACIDS, ACETIC ACIDS, PHTHALLIC ACIDS AND BENZONITRILES

NAMES

2, 4D, AQUA KLEEN, CROTILIN, DEMISE, DIKAMIN, DIKONIRT, EMULSAMINE-E3, ESTERON, FERNOXONE, FORMULA 40, PHENOX, VERTON, WEEDAR, WEEDONE, SOLUTION, PHORDENE, EMBAMINE, FERNIMINE, AGRICORN, CHARDOL, CLOROXONE, SEE, ERBITOX, QUINOXONE, SAVAGE, ENVY, WEEDESTROY, BARRAGE, SALVO, SOLUTION, CRISAMINA

2, 4-dichlorophenoxyacetic acid

TYPE: 2, 4-D is a selective, translocated phenoxy compound used mainly as a postemergence herbicide.

ORIGIN: 1942. Amchem Products Inc. Produced today in the U.S. by Rhone Poulenc, Dow, and others. There are numerous worldwide formulators of this product.

TOXICITY: LD_{50} - 375 mg/kg. May cause eye and skin irritation.

FORMULATIONS:

1. Sodium and ammonium salts—Usually water-soluble. The ammonium salts are rarely found on the market while the sodium salts are marketed for specialized usages.

2. Amine salts—The alkylamines include monomethylene, dimethylamine, isopropylamine, triethylamine, and others. The alkylanolamines include diethanolamine, triethanolamine, mixed isopropanolamines, etc.

3. Highly volatile esters—Methyl, ethyl, butyl, isopropyl, octylamyl, pentyl esters and others containing various Ib. acid equivalent/gal.

4. Low-volatility esters—Contain esters that suppress volatility. Formulations include butoxyethanol, propylene glycol, tetrahydrofurfuryl, propylene gycol butyl ether, butoxy propyl, ethylhexyl, isoctyl and others. These contain various Ib. acid equivalents/gal.

5. Other formulations—Various forms formulated as 10-20% granules, 3-6EC, 6 Ib oil-soluble concentrate/gal and others. Sold mixed with oil solutions, fertilizers, and other pesticides. Often formulated with MCPP, MCPA, 2,4-DP, dicamba and other herbicides.

USES: Asparagus, barley, corn, hay, millets, oats, pasture, rangeland, rice, rye, sorghum, soybeans, grasses grown for seed, fallowland, non crop areas, sugarcane, and wheat. Used also on aquatics, for brush control, as a growth stimulator, and on turf. Used on these and many other crops outside the U.S.

IMPORTANT WEEDS CONTROLLED: Bindweed, Canada thistle, chickweed, cocklebur, goldenrod, ivy, hoary cress, jimsonweed, lambsquarters, locoweed, mustards, pigweed, plantain, Russian thistle, purslane, sunflower, willows, and most other broadleaf weeds.

RATES: Applied at .28-2.3 kg/ha.

APPLICATION: Applied as a postemergence herbicide when the weeds are young and actively growing. The ideal temperature during application should be between 50° and 90°F. On cereals apply when they are fully tillered up to the boot stage. On brush it may be applied as a foliar spray, a basal bark treatment or a stump or frill treatment. In forestry it can be used in site preparation and for conifer release.

PRECAUTIONS: Do not apply where the temperature is over 90°F. Avoid drift.Very susceptible plants include cotton, tomatoes, grapes, fruit trees, and ornamentals. Do not apply near desired plants. Low volatility esters may become volatile at 90°F and above. Excessive 2, 4-D salts in the soil may temporarily inhibit seed germination and plant growth. Application equipment must be thoroughly cleaned with special materials before applying other pesticides to desired crops. Do not use on dichondra, St. Augustine grass, lippagrass, bentgrass or clover turf, or turf that is not yet established. When used as an aquatic herbicide, decaying weeds may off-flavor water for a short period.

ADDITIONAL INFORMATION: Heavy rains will not leach all of the chemical out of the soil. Controls weeds as they germinate when applied preemergence. Delay cultivation as long as possible following application. Amine salts are less hazardous than highly volatile ester formulations to adjoining crops. The shorter the carbon chain, the higher the volatility of the esters. The vapor from low-volatility esters will cause little injury to susceptible crops growing nearby at ordinary temperatures. Esters are used on the same crops as the amines at the same or slightly reduced rates. At the usual dosages, soil microorganisms are not affected. No risk of accumulation in the soil from one year to another. The plants absorb the salt formulations more readily than the acid or ester. Esters tend to resist washing from the plant. Plants are the most susceptible when they are growing rapidly. Susceptible plants usually become malformed before they die. May be mixed with liquid fertilizers and other herbicides.

4

NAMES

DICHLORPROP, 2,4-DP, WEEDONE DPC, REDIPON, DESORMONE, SERITOX, POLYMONE, OPTICA, DUPLOSAN

2-(2,4-dichlorophenoxy) propanoic acid

TYPE: 2,4-DP is a phenoxy compound used as a selective, postemergence herbiclde.

ORIGIN: 1961. The Boots Co. of England, BASF, Rhone-Poulenc and Schering are the principle basic producers today.

TOXICITY: LD_{50} - 500 mg/kg.

FORMULATIONS: 3.7 Ib/gal EC. Formulated as amines and esters. Also formulated with 2, 4-D and other herbicides.

USES: For control of mixed brush on highways, railroads, rangeland, utility right-of-ways and other non crop areas. Also used to control solid stands of post, blackjack, sand shinnery oak, and sandsage. Used outside the U.S. on cereals and other crops.

IMPORTANT WEEDS CONTROLLED: Oaks, pine, fir, spruce, cherry, alder, willow, sandsage, elm, and similar broadleaf species.

RATES: Applied at 2-4 kg ai/ha. May be tank mixed with other brush control herbicides.

APPLICATION: Apply as a foliar spray by ground or air. Apply when foliage is fully developed before it goes dormant. Thorough coverage is necessary.

PRECAUTIONS: Avoid drift. Some regrowth can be expected on some species. Toxic to fish.

ADDITIONAL INFORMATION: May be mixed with oil as well as water. Agitation is required. More effective than 2,4-D on certain broadleaf and woody species.

5

RELATED COMPOUNDS:
1. Dichlorprop-P, Duplosan-DP — A postemergence phenoxy compound developed by BASF for usage outside the U.S. on cereals to control certain broadleaf weeds. It is an isomer of dichlorprop.

NAMES

MCPA-THIOETHYL, PHENOTHIOL, HERBIT, ZERO-ONE, FENOBIT

S-ethyl 2 methyl-4-chlorophenoxy thioacetate

TYPE: MCPA-thioethyl is a phenoxy compound used as a selective, translocated, postemergence herbicide.

ORIGIN: 1970. Hokko Chemical Co. of Japan.

TOXICITY: LD_{50} - 790 mg/kg.

FORMULATIONS: 20% EC. 1.4% granules.

USES: Outside the U.S. on rice, cereals, sugarcane and orchard crops.

IMPORTANT WEEDS CONTROLLED: Sedges, lambsquarter, smartweed, chickweed, bindweed, and other broadleaf weeds.

RATES: Applied at 400-800 g. a.i./ha.

APPLICATION: To rice, apply 7-10 weeks after seeding, when the rice is fully tillered and 15-20 cm above water level. To cereals apply after the late tillering stage.

PRECAUTIONS: Not for sale or use in the U.S. Toxic to fish.

ADDITIONAL INFORMATION: Considered an MCPA analog, but is much safer on rice and other crops. Gives longer control than other phenoxy-type herbicides.

6

NAMES

MCPA, AGRITOX, CHIPTOX, DIKOTEX, HEDONAL-M, KILSEM, KREZONE, LINORMONE, DED-WEED, LINOL, RAPHONE, RHOMENE, RHONOX, TRASAN, WEEDAR MCPA, ZELAN, EMPAL

(4-chloro-2 methylphenoxy) acetic acid

TYPE: MCPA is a postemergence, selective, translocated, phenoxy herbicide.

ORIGIN: 1945. Plant Protection Ltd., England. Produced by a number of basic manufacturers and formulators today.

TOXICITY: LD_{50} - 760 mg/kg. May be irritating to the eyes and skin.

FORMULATIONS: 2EC, 4EC, 20% granules. Formulated as the amine salts, sodium salts and low-volatility esters. Often formulated with other herbicides.

USES: Alfalfa, barley, clover, flax, fallowland, grasses, lespedeza, oats, pastures, peas, timberlands, rangeland, rice, rye, trefoil, turf, vetch, wheat and non crop areas. Used on these and many other crops outside the U.S.

IMPORTANT WEEDS CONTROLLED: Mustards, plantain, arrowhead lily, sedges, bindweed, burhead, lambsquarters, puncturevine, ragweed, cocklebur, purslane, peppergrass, and many other broadleaf weeds.

RATES: Applied at .2-2 lb actual/A.

APPLICATION: Use on young, rapidly growing weeds. Use higher rates on perennials. Apply to cereals after tillering but before they reach the boot stage. To peas apply after they have reached the 3 node stage but before flowering.

PRECAUTIONS: Do not apply when the temperature is below 40°F or above 90°F. Do not apply to grains in the boot or dough stage. Do not contaminate irrigation water. Very sensitive crops include grapes, vegetables, fruit trees, ornamentals, and cotton. Avoid drift. In certain areas, injury has resulted on bent, buffalo, carpet, and St. Augustine grasses. Corrosive to aluminum and zinc.

ADDITIONAL INFORMATION: Safer on crops than 2,4-D. More effective on some broadleaf weeds than 2,4-D, but considered less effective on most broadleaf weeds. May be used with other herbicides.

NAMES

MCPB, BEXANE, CAN-TROL, LEGUMEX, BELLMAC, THISTROL, TRITROL, TROPOTOX, TROTOX

4(4-chloro-2-methylphenoxy) butanoic acid

TYPE: MCPB is a phenoxy compound used as a selective, postemergence, translocated herbicide.

ORIGIN: 1955. May and Baker Co. of England. Rhone Poulenc is the principle basic producer.

TOXICITY: LD_{50} - 680 mg/kg.

FORMULATIONS: 2 EC (sodium salt). Formulated with other herbicides.

USES: Peas. Outside the U.S. it is used on peas, cereals and pastures.

IMPORTANT WEEDS CONTROLLED: Pigweed, purslane, lambsquarters, smart-weed, mustards, ragweed, dock, thistles, Canada thistle, buttercups, and many other annual broadleaf weeds.

RATES: Applied at 1.7-3.4 kg a.i./ha.

APPLICATION: Apply to peas when the crop has 6-12 nodes but before flowering. Weeds should be less than 3 inches tall.

PRECAUTIONS: Do not apply when temperatures exceed 90°F. Do not apply during hot weather or drought conditions. Avoid drift.

ADDITIONAL INFORMATION: Susceptible weeds convert this compound into MCPA and are killed, whereas most legumes are unable to make this conversion and are tolerant. More selective than 2,4-D and MCPA in many instances and may be used with greater safety. Injury in the form of twisting may occur with some pea varieties.

NAMES

MECOPROP, MCPP, CLOVOTOX, COMPITOX,
CMPP, CORNOX-PLUS, ISO-CORNOX, HEDONAL-MCPP, KILPROP, LIRANOX, MECOMEC, PROPAL, MECOPEX, MEPRO, RUNCATEX, DUPLOSAN-KV, ASTIX, OPTICA, CLENECORN, HERRIFEX, HYMEC

2-(4-chloro-2-methyl phenoxy) propanoic acid

TYPE: MCPP is a phenoxy compound used as a selective, translocated compound applied as a postemergence herbicide.

ORIGIN: 1953. Boots Pure Drug Company of England. Schering AG of Germany is one of the basic producers. Marketed by a number of formulators.

TOXICITY: LD_{50} - 558 mg/kg. May cause eye irritation.

FORMULATIONS: 2.5EC, 4EC. Available as the amine, sodium and potassium salts. Formulated with a number of postemergence herbicides.

USES: Turf and non- crop areas. Used outside the U.S. on cereals, turf and forage grasses.

IMPORTANT WEEDS CONTROLLED: Clovers, chickweed, knotweed, ragweed, shepherd's purse, pigweed, mustard, lambsquarters, cleavers, ground ivy, plantain, and other broadleaf weeds.

RATES: Apply at 1.8-3.9 kg a.i./ha.

APPLICATION: Apply uniformly and thoroughly, but not to the point of runoff. Apply when weeds are growing vigorously, but before they become established. A second treatment may be made for the control of certain hard-to-kill weeds. Avoid cutting or mowing for 2-3 days before and after treatment.

PRECAUTIONS: Do not apply when turf is suffering for water. Do not apply in unusually wet or hot weather. Avoid drift. Do not apply when temperatures exceed 90°F. Do not use the first cutting after treatment as a mulch for flowers and vegetables.

ADDITIONAL INFORMATION: Safer than other phenoxy compounds on fine turf grasses (bentgrass, bluegrass, etc.). May be mixed with 2,4-D and other herbicides to control a larger range of weeds. Relatively slow in its action, requiring 3-4 weeks for control. Effective only on broadleaves. May be applied close to woody shrubs bordering turf without injury if foliage is not contacted. Non-corrosive. Weeds such as black medic, dock, English daisy, knotweed, mallow, purslane, and oxalis should be treated early for best results. Used mainly in combination with other herbicides.

RELATED MIXTURES:

1. TRIMEC—Combinations of MCPP, dicamba and 2, 4-D marketed by PBI/Gordon as a selective herbicide on turf.

NAMES

2,4-DB, BUTYRAC, BUTOXONE, EMBUTOX, EMBUTONE

4-(2,4-dichlorophenoxy) butanoic acid

TYPE: 2,4-DB is a phenoxy compound used as a selective, postemergence, translocated herbicide.

ORIGIN: 1947. May and Baker Inc. of England. Produced by Rhone Poulenc and Cedar Chemical Co. in the U.S.

TOXICITY: LD_{50} - 700 mg/kg. May be irritating to the eyes and skin.

FORMULATIONS: 1.75, 2EC. Formulated as the amine salts and esters. The amine salts are soluble in water, while the esters are soluble in oil and emulsifiable in water.

USES: Alfalfa, clovers, peanuts, soybean and trefoil. Used on these and other crops outside the U.S.

IMPORTANT WEEDS CONTROLLED: Mustards, cocklebur, shepherd's purse, lambsquarters, pigweeds, ragweed, bindweed, plantain, nightshade, and many other broadleaf weeds. Not effective on grasses or established broadleaves.

RATES: Applied at .5-2 lb actual/A.

APPLICATION: Used as either a band or broadcast treatment. Apply after the crop has started growth and when the weeds are small. The weeds should be less than 3 inches high for the most effective control. Applied to either seedlings or established legumes. There may be some temporary injury to the crop but it will soon outgrow this without the final yields being affected. To control cocklebur In soybeans, apply overall 7-10 days prior to bloom and up to midbloom. Later applications may affect flowers, reducing yields. Treated soybeans will show a slight twisting, but they will outgrow this in 3-5 days. A postemergence, directed spray may also be made when the soybeans are 8-12 inches tall. Soybeans should be growing actively at the time of application. Nurse crops of barley, oats, and wheat underseeded to legumes may be treated when the nurse crop is 1-6 inches tall. Legumes should be in the 1-2 trifoliate leaf stage before application.

PRECAUTIONS: Do not treat if temperatures are expected to go over 90°F or below 40°F in the following few days. Sweet clover should not be treated. Mature weeds will not be effectively controlled. Do not apply preemergence. Treat before legumes show flowers. Avoid drift. Do not treat soybeans infested with phytophthora, as injury will result. Do not treat soybeans with a tank mix of carbaryl insecticide or severe injury will result. Do not use on peas. On seedling alfalfa, delay irrigation for at least 10 days following application.

ADDITIONAL INFORMATION: Legumes convert this compound to 2, 4-D very slowly, so they are resistant, while most weeds convert it very fast, making them highly susceptible. Ester formulations are faster acting and give a better kill of hard-to-control species. Non-corrosive. Absorbed through the foliage only. Weeds should be controlled within 3 weeks of treatment. May be tank-mixed with other herbicides.

NAMES

BIFENOX, MODOWN

Methyl 5-(2,4-dichlorophenoxy)-2-nitrobenzoate

TYPE: Bifenox is a diphenyl ether compound used as a selective, preemergence herbicide.

ORIGIN: 1970. Mobil Chemical Company. Now being marketed by Rhone Poulenc.

TOXICITY: LD_{50} - 6400 mg/kg.

FORMULATIONS: 2EC, 80% WP. 4 Ib/gal. flowable, 7 and 10% granules. Sold in some countries formulated with other herbicides.

USES: Outside the U.S. on corn, rice, sorghum, barley, oats, wheat, and soybeans.

IMPORTANT WEEDS CONTROLLED: Barnyardgrass, jimsonweed, kochia, lambsquarters, nightshade, pigweed, smartweed, velvetleaf and others.

RATES: Applied at .75-1 kg a.i./ha.

APPLICATION: Make application after planting, but before crop seedlings emerge. Used on cereals postemergence. On rice, it is applied both preemergence or postemergence up to the 2 leaf stage.

PRECAUTIONS: No longer used in the U.S. Heavy rainfall after application, especially one that splashes treated soil onto the plants, may cause crop injury. This injury may not appear for 2-3 weeks, and the crop will grow out of it. Do not incorporate this material. Toxic to fish. Store above 32°F.

ADDITIONAL INFORMATION: Agitate while applying. Forms a chemical barrier that stops the germinating of weeds as they attempt to grow through the surface.

Primarily a broadleaf herbicide. Water solubility is .35 ppm. Mixed with other herbicides to increase their spectrum of control.

BENZOIC, ACETIC ACIDS
AND
PHTHALLIC COMPOUNDS

NAMES

CHLORAMBEN, AMIBEN

Cl

NH$_2$ Cl

C=O
OH

3-amino-2,5-dichlorobenzoic acid.

TYPE: Chloramben is a benzoic acid compound used as a selective, preemergence herbicide.

ORIGIN: 1958. Union Carbide.Being marketed by Rhone Poulenc.

TOXICITY: LD$_{50}$ - 7150 mg/kg. May cause skin irritation.

FORMULATIONS: 25C, 83% WDG, 10% granules.

USES: Outside the U.S. on asparagus, corn, beans, cucurbits, peanuts, soybeans, sunflowers, sweet potatoes, and ornamentals.

IMPORTANT WEEDS CONTROLLED: Crabgrass, foxtail, chickweed, coffeeweed, teaweed, lambsquarters, kochia, pigweed, ragweed, smartweed, mustard, velvetleaf, and many other annual broadleaves and grasses.

RATES: Applied at 2-4 Ib actual/A.

APPLICATION: Applied as a preemergence treatment. Rainfall or irrigation is required to move it into the soil.

PRECAUTIONS: Emerged weeds are not controlled. Avoid drift. No longer for sale or use in the U.S.

ADDITIONAL INFORMATION: Effective on both grasses and broadleaves. Leaves no soil residue to injure the following crop. Requires 1/2 inch of rain or sprinkler irrigation within 10-14 days to carry it into the soil to make it effective. Morningglory, jimsonweed, buffalobur, and cocklebur are not consistently controlled. Perennial weeds are not controlled. Non-volatile. Use higher rates on heavy clay soils. Unlike most

17

preemergence herbicides, the activity is nearly the same on muck soils as on mineral soils. Soil incorporation of 2-6 inches decreased weed control, probably due to the dilution of the chemical. Non-corrosive. Inhibits root development of seeding weeds. Control should last 6-8 weeks.

NAMES

DICAMBA, BANEX, BANVEL, TRACKER, MEDIBEN, DATAMIN, DIANAT, DYVEL, TROOPER

3,6-dichloro-2-methoxybenzoic acid

TYPE: Dicamba is a benzoic acid derivative used as a pre and postemergence, selective, translocated herbicide.

ORIGIN: 1965. Sandoz Agro.

TOXICITY: LD_{50}- 1040 mg/kg.

FORMULATIONS: 4EC, 2EC. Available as the dimethylamine salt and as the sodium salt. Also available formulated with other herbicides.

USES: Barley, asparagus, corn, oats, sorghum, grasses (raised for seed), aquatics, non-crop areas, pastures, rangelands, sugarcane, turf, and wheat.

IMPORTANT WEEDS CONTROLLED: Sheep sorrel, chickweed, knotweed, clovers, cockles, lambsquarters, pigweed, velvetleaf, Canada thistle, Russian knapweed, aquatic weeds, cactus, bindweed, dock, and most other broadleaf annuals and perennials.

RATES: Applied at .25-8 Ib actual/A.

APPLICATION: Applied as a postemergence herbicide when the weeds are young and actively growing. On small grains apply from the 2-5 leaf stage up to the early boot stage. On corn apply before it is 36 inches tall. For control of woody plants apply as a foliar treatment, a basal spray and a cut stump or frill treatment.

18

PRECAUTIONS: Avoid drift. Soybeans and dry beans are very susceptible to this chemical. Do not apply in the drip line of desired trees. Do not use on com grown on sandy soils, low in organic matter. Do not apply by air if sensitive crops are near.

ADDITIONAL INFORMATION: Compatible with most other pesticides. Residues disappears within several weeks or months, depending upon the weather. Relatively mobile in the soil. May be mixed with a number of other herbicides.

RELATED MIXTURES:

I . WEEDMASTER—A pasture and non-cropland herbicide containing 1 Ib dicamba and 2.87 Ib. 2, 4-D. Marketed by Sandoz Agro.

2. MARKSMAN—A combination of 1.1 lb. dicamba and 2.1 lb. atrazine developed by Sandoz Agro for corn, grain, sorghum and fallow systems.

3. BANVEL 520, BANVEL 720—A combination of dicamba and 2,4-D developed by Sandoz Agro to use for woody species and broadleaf weed control in industrial non-crop areas.

NAMES

BENAZOLIN, **BENASALOX, CHAMILOX,**
ASSET, TILLOX, LEGUMEX EXTRA, CRESOPUR, GALTAK, KEROPUR

4-chloro-2-oxobenzothiazolin-3-ylacetic acid

TYPE: Benazolin is a benzoic acid compound applied as a selective, postemergence herbicide.

ORIGIN: 1965. The Boots Co. of England. Now marketed by Schering Agrochemicals of Germany.

TOXICITY: LD_{50} - 4800 mg/kg. May irritate eyes and skin.

FORMULATIONS: Potassium and sodium salt solutions, or as an ester . Formulated with other herbicides.

USES: Used outside the U.S . on cereals, clovers, alfalfa, soybeans, corn, flax, peas, and rape.

IMPORTANT WEEDS CONTROLLED: Chickweed, cocklebur, lambsquarters, night-shade, pigweed, sesbania, sunflower, velvetleaf, veronica spp, cleavers and other broadleaf weeds.

RATES: Applied at .25-.375 Ib ai/A.

APPLICATION: Applied postemergence when the weeds are actively growing. There is a very high degree of selectivity in cereals and clover. Usually mixed with other herbicides to give a broader spectrum of weed control. Apply to cereals anytime between the appearance of the first leaf and lhe onset of jointing. May be applied with a crop oil concentrate. Applied to soybeans when they are in the 2-3 trifoliate leaf stage or later.

PRECAUTIONS: Some injury to soybeans has occurred but the crop quickly outgrows it. Do not use in hard water. Mildly corrosive to some metals. Toxic to fish. Not for sale or use in the U.S.

ADDITIONAL INFORMATION: Low temperatures do not reduce the effectiveness. Only very slight preemergence activity. The mode of action is similar to mecoprop only usually slower. Shows high synergistic activity when mixed with dicamba. A foliar-absorbed herbicide which is translocated throughout lhe plant. A growth regulator at low rates for promoting plant growth.

RELATED COMPOUNDS:

TCA—A preemergence herbicide used outside the U.S. on sugar beets. It was first introduced in 1946 by Dow Chemical Co. Hoechst is now one of the basic producers.

Dalapon, Dowpon, Basapon—An older compound no longer used in the U.S. Used in some parts of the world on noncrop areas and plantation crops for the control of perennial grasses. BASF is one of the basic producers.

NAMES

NAPTALAM, NPA, ALANAP

(sodium 2-[(1-naphthalenylamino) carbonyl] benzoic acid)

TYPE: Naptalam is a selective, preemergence, phthalic-acid herbicide.

ORIGIN: 1949. Uniroyal Chemical.

TOXICITY: LD_{50} - 1770 mg/kg. Causes eye irritation.

FORMULATIONS: 2EC. Sold as the sodium salt.

USES: Peanuts and cucurbits. Outside the U.S. it is used on these as well as soybeans, ornamentals and other crops.

IMPORTANT WEEDS CONTROLLED:

Pigweed, lambsquarters, ragweed, chickweed, shepherd's purse, jimsonweed, purslane, velvetleaf, mustard, cocklebur, groundcherry, and other broadleaf weeds.

RATES: Applied at 2-5.5 kg a.i./ha.

APPLICATION: Applied as a preemergence or postemergence application. Rainfall or irrigation is necessary to take it into the soil.

PRECAUTIONS: Emerged weeds will not be controlled. May be detrimental to the crop when applied to soils low in organic matter, low in clay content, or with an extremely high pH (9-12). Do not apply when temperature is above 100°F and do not use in wooden spray tanks. Do not use on extremely high peat or muck soils. Beets, tomatoes, spinach, and lettuce are extremely sensitive to this chemical. Avoid application to the foliage of nursery plantings. Do not apply to emerged soybeans. Avoid freezing in storage.

21

ADDITIONAL INFORMATION: A non-volatile chemical activated by moisture. Effective for 3-8 weeks. Completely decomposes in 6-8 weeks. Does not corrode equipment and is easily cleaned by water alone. Moves both vertically and laterally in the soil. Acts on the seed before, and as it germinates. May be absorbed through the root system after germination. Does not kill by contact so usually not effective on established weeds. May be tank mixed with other herbicides.

NAMES

QUINOCLAMIN, ACN, MOGENTON, OGK

$C_{10}H_6O_2NCl$

2-amino-3-chloro-1,4 napthoquinone

TYPE: Quinoclamin is a quinone compound used as a selective herbicide-algicide.

ORIGIN: Agro-Kanesho Co. Ltd. of Japan 1975.

TOXICITY: LD_{50} 1360 mg/kg.

FORMULATION: 9% granules.

USES: Outside the U.S. on rice and for greenhouse and turf moss control.

IMPORTANT WEEDS CONTROLLED: Mosses, floating aquatic weeds, algae, Sagittaria spp, Potamogeton spp. and others.

RATES: Applied at 2.7-3.6 kg a.i./ha.

APPLICATION: Applied as a granule material to the water surface. Stop the flow of water after treatment.

PRECAUTION: Not for sale or use in the U.S. Toxic to fish.

ADDITIONAL INFORMATION: Compatible with other herbicides. Contact activity under submerged conditions. Fast acting. Rice is tolerant to this material. Retards photosynthesis. Water is needed to activate the material.

22

NAMES

ENDOTHALL, AQUATHOL, ENDOTHAL, HYDROTHAL-47, HYDROTHAL-191, DES-I-CATE, ACCELERATE, HYTROL, HERBICIDE 273

7-oxabicyclo(2,2,1)heptane-2,3-dicarboxylic acid

TYPE: Endothall is a selective pre and postemergence, carboxylic-acid herbicide.

ORIGIN: 1951. Pennwalt Corp.

TOXICITY: LD_{50} - 38 mg/kg. Irritating to eyes, nose, skin, and throat. Aquatic formulations (inorganic salts) are safe to fish in 100-500 ppm concentrations.

FORMULATIONS: Formulated as a number of amine salts and as the sodium and potassium salt in both liquid and granular form.

USES: Sugar beets. Used as a desiccant on alfalfa, potatoes, clover, and cotton. Used as an aquatic herbicide and as a turf herbicide.

IMPORTANT WEEDS CONTROLLED: Cheat, pigweed, kochia, foxtail, bur clover, barley, crabgrass, annual bluegrass, ragweed, purslane, shepherd's purse, aquatic weeds, and many others.

RATES: Applied at 1-6 Ib actual/A.

APPLICATION: Applied as a preemergence herbicide. Rainfall is required to move it into the root zone. As a postemergence application on sugar beets apply when they are in the 4-6 leaf stage. As a preharvest desiccant apply 7-14 days prior to harvest. As an aquatic herbicide apply to the water surface or inject it into the water to control submerged aquatics and some algae species. Water temperature should be 65° F or above.

23

PRECAUTIONS: Toxic to certain fish. Do not apply postemergence to sugar beets if the temperature exceeds 85° F.

ADDITIONAL INFORMATION: Kills on contact. Water should be 65°F or above for best results on aquatic weeds. Hydrothal formulations are more toxic lo fish than the Aquatrol formulations. Often mixed with other herbicides on sugar beets for grass control. On cotton, as a preharvest defoliant, tank mix with chlorate or phosphate defoliants.

NAMES

CHLOROTHAL, DCPA, DACTHAL, DACTHALOR

Dimethyl 2,3,5,6-tetrachloro-1,4-benzenedicarboxylate

TYPE: DCPA is a benzoic-acid compound, used as a selective, preemergence herbicide.

ORIGIN: 1960. Diamond Alkali Co. Now marketed by ISK Biotech.

TOXICITY: LD_{50} - 3000 mg/kg. May cause slight eye irritation.

FORMULATIONS: 75% WP, 5% granules, 6 Ib/gal flowable.

USES: Broccoli, brussels sprouts, cabbage, cantaloupes, cauliflower, collards, cotton, cucumbers, dry beans, eggplant, garlic, honeydew melons, horseradish, kale, lettuce, mung beans, mustard greens, nursery stocks, onions, ornamentals, peppers, potatoes, snap beans, southern peas, squash, strawberries, sweet potatoes, tomatoes, turf, turnips, watermelons, and yams.

IMPORTANT WEEDS CONTROLLED: Crabgrass, foxtails, barnyardgrass, goosegrass, bluegrass, lambsquarters, purslane, ground cherry, chickweed, pigweed, spurge, dodder, and others.

RATES: Applied at 6-10.5 lb actual/A.

APPLICATION: Applied as a preemergence herbicide. Incorporation lessens the effectiveness except where crops are irrigated; then a marked increase in activity is noted. Agitate while spraying. Eliminate any established weeds before treatment. On some crops, apply 4-6 weeks after seeding.

PRECAUTIONS: Apply only to mineral soils. Susceptible crops include: beets, spinach, trefoil, lespedeza, and flax. Injury has been reported on bentgrass and red fescue turf as well as dichondra. Do not use on lima beans. Perennial weeds, mustard, ragweed, smartweed, and velvetleaf are among the weeds not controlled. Do not use on putting greens.

ADDITIONAL INFORMATION: Widely used for crabgrass control in turf. Controls germinating seeds but has little effect when applied postemergence. Lawns must not be reseeded for 60 days after treatment. Not effective on muck and peat soils. Season-long control may be expected.

NAMES

DICHLOBENIL, **BARRIER, CASORON, DYCLOMEC, NOROSAC, SILBENIL**

2,6-dichlorobenzonitrile

TYPE: Dichlobenil is a benzonitrile compound used as a preemergence,selective herbicide.

ORIGIN: 1960. N. V. Philips-Duphar, in Holland. Now marketed by PBI Gordon Corp., Uniroyal and Shell Chemical Co.

TOXICITY: LD_{50} - 2126 mg/kg.

ORMULATIONS: 4% and 10% granules, 50 WP.

USES: Apples, grapes, cranberries, cranberries, blueberries, filberts, cherries, nectarnes, peaches, pears, prunes, plums, ornamentals, and non-crop areas. Used for aquatic weed control and total vegetation control. Used in some countries on transplant rice.

IMPORTANT WEEDS CONTROLLED: Horsetail, smartweed, shepherd's purse, nutsedge, plantain, rushes, barnyardgrass, pondweeds, naiad, pigweed, lambsquarters, crabgrass, foxtails, dandelion, purslane, quackgrass, dodder, and others.

RATES: Applied at 1.5-8 actual Ib/A.

APPLICATION:

1. Aquatic Weeds—Treat in the very late fall or early spring before emergence of weeds. Used in lakes, ponds, and reservoirs. Do not use water for irrigation, livestock, or human consumption. Do not use fish from treated waters for 90 days after application. Do not use in shellfish waters.

2. Orchards—Apply in a weed-free situation before the seeds germinate. If the soil is dry, shallow incorporation is effective. Apply to established trees.

3. Noncrop Areas—May be used on ornamentals, wind breaks, forestry park lands, under asphalt, and as a soil sterilant giving control for the entire season.

PRECAUTIONS: Do not use for 4 weeks after transplanting nursery stock. Do not use on light, sandy soils. Do not apply during period of high soil temperature. Do not use in greenhouses.

ADDITIONAL INFORMATION: Incorporation increases the effectiveness of this compound, although it may decrease the tolerance of some slightly susceptible plants. Deep-rooted plants are tolerant. Inhibits the sprouting of potatoes. Control may last for the entire season. Does not build up in the soil. No contact activity. Does not move in the soil to any great extent. Volatile to the extent that there is a partial loss of chemical from the soil surface, especially under high-temperature conditions. When applied to existing weeds, control may take 2-3 weeks.

NAMES

BROMOXYNIL, BROMINAL, BUCTRIL, PARDNER, TORCH, MERIT, LITAROL-M, BROMINEX, KORIL, KORILENE, SABRE, TRON

3,5-dibromo-4-hydroxybenzonitrile

TYPE: Bromoxynil is a hydroxybenzonitrile compound used as a selective, contact, postemergence herbicide.

ORIGIN: 1963. In England by May-Baker, Ltd. and in the U.S. by Amchem Products. Rhone Poulenc is the principle basic producer today.

TOXICITY: LD_{50}- 190 mg/kg.

FORMULATIONS: 2EC, 4EC. Formulated as different salts and esters. May be formulated with other herbicides.

USES: Wheat, alfalfa, oats, rye, flax, garlic, onions, mint, corn, sorghum, barley and non crop areas. Experimentally being used on cotton, soybeans, peanuts, rice and sugarcane.

IMPORTANT WEEDS CONTROLLED: Fiddleneck, dog fennel, tarweed, smartweed, buckwheat, mustards, lambsquarters, London rocket, shepherd's purse, nightshade, groundsel, and others.

RATES: Applied at 460-600 g a.i./ha.

APPLICATION: Apply when cereals are in the 2-3-leaf stage up to the boot stage and before weeds are past the 3-4-leaf stage. On older weeds, apply at a higher rate but before flower formation. Do not apply to crops stressed from lack of moisture. Used to control Russian thistle and other hard-to-kill weeds on railroads and other non-crop areas. Use on corn from the 3-leaf stage until it is 20 inches tall. Apply to alfalfa in the 2-4 trifoliate leaf stage.

PRECAUTIONS: Do not apply to cereals during or after the boot stage. Established eeds will not be controlled. Avoid drift. Toxic to fish. Filaree and chickweed to this compound. Do not treat flax in the bud stage or in weather over 85°F. ; acid ester formulation can no longer be used in the U.S.

B

)NAL INFORMATION: Non-flammable and non-corrosive. Absorbed by the ,ut not translocated. Inhibits photosynthesis and plant respiration. More active eeds are growing rapidly. Controls many weeds not easily controlled by 2,4-D. : combined with 2,4-D, MCPA, or other phenoxy-type herbicides. Can be applied / young grasses without injury, unlike the phenoxy compounds. Non-volatile. May ,plied with liquid fertilizers.

LATED MIXTURES:

. Buctril + Atrazine — A combination of bromoxynil and atrazine marketed by Rhone ?oulenc to control broadleaf weeds in corn and sorghum.

2. Bronate — A combination of bromoxynil and MCPA marketed by Rhone Poulenc to control broadleaf weeds in cereals.

NAMES

IOXYNIL, ACTRIL, BANTROL, CERTROL, IOTOX, MATE, TOTRIL, ACTRILAWN, IOTRIL, AXALL, STEXAL

4-hydroxy-3,5-diiodobenzonitrile

TYPE: loxynil is a hydroxybenzonitrile compound used as a selective, contact, postemergence herbicide.

ORIGIN: 1963. May and Baker of England and Amchem Products in the U.S. Rhone Poulenc is the principle basic producer today.

TOXICITY: LD_{50} - 110 mg/kg. May cause eye irritation.

FORMULATIONS: 25%EC. Also sold mixed with other postemergence herbicic

IMPORTANT WEEDS CONTROLLED: Dog fennel, smartweed, buckwheat, chi weed, speedwell, dead nettle, and other broadleaf weeds.

USES: Used in many countries on wheat, sugarcane, oats, barley, rye, onions, and tu.

RATES: Applied at 400-600 g a.i./ha.

APPLICATION: Apply to weeds in the seedling stage.

PRECAUTIONS: Not registered for use in the U.S. Do not spray on or near water. Toxic to fish.

ADDITIONAL INFORMATION: Absorbed by the foliage, but only translocated. Slightly non-volatile. No residual activity. Mixed with other compounds to give broader-spectrum control.

s.

k-

f.

DINTRO ANALINES NITRITES, AMIDES, ACETAMIDES AND ANILIDES

NAMES

TRIFLURALIN, CRISALINA, DIGERMIN, ELANCOLAN, HERITAGE, IPESSAN, RIVAL, TRI-4, TRUST, TRIGARD, TREFLAN, TREFLANOCIDE, TRIDENT, TRIFLUREX, TRILIN, TRISTAR, TRIFLURALINA

$$CH_3 — CH_2 — CH_2 \diagdown \quad \diagup CH_2 — CH_2 — CH_3$$
$$N$$
$$NO_2 \quad NO_2$$
$$F — C — F$$
$$F$$

2,6-dinitro-N,N-dipropyl-4-(trifluoromethyl) benzenamine

TYPE: Trifluralin is a dinitroanaline compound used as a selective, preplant herbicide.

ORIGIN: 1959. Elanco Products Company, now Dow Elanco. Also being produced by other manufacturers.

TOXICITY: LD_{50} - 3700 mg/kg

FORMULATIONS: 4EC, 5% and 10% granules, 5EC. Formulated with other herbicides.

USES: Alfalfa, almonds, apricots, asparagus, broccoli, Brussels sprouts, cabbage, cantaloupes, carrots, castorbeans, cauliflower, celery, cherries, citrus, collards, cotton, corn, cucumbers, dry beans, flax, garlic, guar, hops, grapes, kale, lentils, lima beans, mung beans, mustard, okra, ornamentals, peaches, peas, pecans, peppers, plums, potatoes, prunes, rape, safflower, snapbeans, peanuts, southern peas, sugar beets, sugarcane, sunflower, tomatoes, turnip greens, walnuts, watercress, wheat, and watermelons. Used on these and other crops outside the U.S.

IMPORTANT WEEDS CONTROLLED: Barnyardgrass, cheat, chickweed, crabgrass, foxtails, panicum, goosegrass, shattercane, johnsongrass, lambsquarters, pigweed, puncture vine, purslane, Russian thistle, sandbur, sprangletop, stinkgrass, and many others.

RATES: Applied at .5-2 Ib actual/A. Use the lower rates on sandy soils.

APPLICATION:

Incorporated into the soil 2-4 inches deep, within 24 hours after application. Common soil incorporation equipment, from double discing to power and ground-driven rotary tillers, has proven most effective. Preplant incorporation is the most commonly used method, but a postplant incorporation may be accomplished for use on a number of crops. Field cultivators may also be used for incorporation.

PRECAUTIONS: Incorporate within 24 hours to prevent loss of activity. Do not allow to freeze. Tolerant weeds include velvetleaf, nightshade, jimsonweed, buffalobur, horsenettle, nutgrass, cocklebur, and established annual and perennial weeds. Any condition which places a stress on the crop may increase the possibility of crop injury from this material. Carryover, causing injury to some crops, may occur in certain areas.

ADDITIONAL INFORMATION: Kills weed seeds as they germinate. Rainfall is not required to activate the chemical. Soil incorporation gives greatest effectiveness. Absorbed by the soil and is extremely resistant to leaching. Little, if any, lateral movement in the soil. Full-season weed control can be expected. Breaking a crust or subsequent cultivation will not reduce the effectiveness and may actually increase it due to additional soil incorporation. Effective on peat or muck soils up to 20% in organic matter. May be combined with both dry and liquid fertilizer. May be applied by air. May be mixed with other herbicides.

RELATED MIXTURES:

1. SALUTE—A combination of trifluralin and metribuzin developed by Miles Inc. to be used, preplant incorporated, on soybeans.

2. COMMENCE—A combination of trifluralin and clomazone used as a preplant herbicide on soybeans, marketed by FMC and DowElanco.

RELATED COMPOUNDS:

1. FLUCHLOR*ALIN*, BASALIN—A dinitroanaline preplant incorporated herbicide developed by BASF for usage on cotton, peanuts, potatoes, rice, jute, soybeans and sunflowers. Not marketed in the U.S.

2. *PROFLURALIN*, TOLBAN, PREGARD—A dinitroanaline compound introduced by Ciba Geigy in 1973 and sold in Europe on cotton, soybeans and other crops as a preplant incorporated herbicide.

3. *DINITRAMINE*, COBEX—A dinitroanaline compound first introduced by U.S.

Borax as a preplant soil incorporated herbicide. It is now produced by Wacker Chemie of Germany and sold outside the U.S. for usage on beans, carrots, cotton, peanuts, soybeans, sunflowers, peppers, tomatoes and cole crops.

4. *BUTRALIN*, AMEX, TAMEX, AMEXINE—A dinitroanaline compound introduced by Amchem Prod. Co. and sold outside the U.S. by Rhone Poulenc as a preplant incorporated herbicides on cotton and soybeans. It is also used as a growth regulator for sucker control in tobacco outside the U.S.

5. *FLUMETRALIN*, PREMIER—A dinitroanaline compound being evaluated by Ciba Geigy for use on turf and ornamentals as a preemergence treatment. It is currently sold on tobacco for sucker control under the tradename Prime +.

6. *ISOPROPALIN*, PAARLAN—A dinitroanaline compound developed by DowElanco to use on transplant tobacco. It is no longer being marketed in the U.S.

NAMES

BENEFIN, BENFLURALIN, BALAN, BALFIN, BENEFEX, FLURAL, BONALAN, EMBLEM, FLUBALEX, QUILAN, BANAFINE

N-butyl-N-ethyl-2,6-dinitro-4-(trifluoromethyl) benzenamine

TYPE: Benefin is a dinitroanaline compound used as a selective, preplant herbicide.

ORIGIN: 1966. Elanco Products Company, now Dow Elanco.

TOXICITY: LD$_{50}$ 10,000mg/kg.

FORMULATIONS: 1.5EC, 2.5% granules. 60% DF.

USES: Established turf, lettuce, tobacco, peanuts, alfalfa, clovers, and trefoil. Used on these and other crops outside the U.S.

IMPORTANT WEEDS CONTROLLED: Crabgrass, barnyardgrass, junglerice, goosegrass, panicum, pigweed, lambsquarters, purslane, carpetweed, and a number of others.

RATES: Applied at .75-1.5 Ib actual/A in 20-40 gal of water.

APPLICATION: Applied as a preplant treatment up to 10 weeks prior to planting. Incorporate into the soil immediately. Incorporation may be done with PTO driven equipment, double discs, rolling cultivators, etc. Incorporators should be set to cut 4-6 inches deep and travel at 4-6 miles/hour. To turf, apply and water in immediately.

PRECAUTIONS: Incorporation less than 2 inches deep may result in erratic weed control. Do not use on muck or peat soils. Do not allow to freeze. Do not plant wheat, rye, grass crops, or onions within 10 months of the application. Milo, corn, oats, beets, and spinach should not be planted for 12 months. Tolerant weeds include nightshade, mallows, nutgrass, cocklebur, groundsel, and ragweed.

ADDITIONAL INFORMATION: Season-long control can be expected. Kills weed seeds as they germinate. An analog of trifluralin. Rainfall is not required to activate the chemical. Shallow cultivations will not reduce the effectiveness. Used for crabgrass control in established turf grass.

RELATED MIXTURES:

1. TEAM—A granular combination of benefin and trifluralin developed by DowElanco for use on turf.

2. XL—A granular combination of benefin and oryzalin developed by DowElanco for use on turf.

3. REGALSTAR—A combination on fertilizer of benefin and oxadiazon developed by Regal Chemical Co. for use on turf.

NAMES

ORYZALIN, DIRIMAL, RYZELAN, SURFLAN

$$CH_3 - CH_2 - CH_2 \quad CH_2 - CH_2 - CH_3$$

$$N$$

$$NO_2 \qquad\qquad NO_2$$

$$O - S - O$$

$$NH_2$$

4-(dipropylamino)-3,5-dinitrobenzenesulfonamine

TYPE: Oryzalin is a dinitroanaline compound used as a selective, preemergence and preplant herbicide.

ORIGIN: 1968. Elanco Products Company now DowElanco.

TOXICITY: LD_{50} - 10,000 mg/kg. May cause skin irritation.

FORMULATIONS: 4 Ib/gal flowable. 75 WP.

IMPORTANT WEEDS CONTROLLED: Barnyardgrass, crabgrass, foxtails, goosegrass, johnsongrass (from seed), fall panicum, brachiaria, crowfootgrass, purslane, carpetweed, Florida pusley, lambsquarters, pigweed, and others.

USES: Turf, bearing and non-bearing trees and vines, and ornamentals. May be used at high rates for non-cropland weed control. Used outside the U.S. on cotton, peanuts, rape, soybeans, sunflowers and a number of other crops.

RATES: Applied at .75-2 Ib actual/A.

APPLICATION: Applied as a preemergence, non-incorporated treatment, either before, during, or after planting. Rainfall or overhead irrigation of .5 inch is required to move this material into the root zone of the germinating weeds. May be applied through sprinkler systems. Used only on established warm season turf.

PRECAUTIONS: Toxic to fish. Do not use on soils containing more than 5% organic matter. Over application may result in crop injury. Do not plant any root crop for 12 months from the time of application.

ADDITIONAL INFORMATION: Partial control of velvetleaf, smartweed, spurge, nightshade, momingglory, teaweed, and ragweed has been obtained. Not readily decomposed by sunlight. Water solubility is 2.5 ppm. Not volatile when applied to the soil surface. Season-long control may be expected. Used in combination with other herbicides to increase the weed spectrum. Does not control established weeds. May be mixed with liquid fertilizers or impregnated on dry bulk fertilizer.

NAMES

PRODIAMINE, BARRICADE, BLOCKADE, ENDURANCE, MARATHON, RYDEX, KUSABLOCK, SENTINEL

N^3,N^3-di-N-propyl-2,4-dinitro-6(trifluoromethyl)-m-phenylenediamine

TYPE: Prodiamine is a dinitroanaline compound used as a selective preplant or preemergence herbicide.

ORIGIN: U.S. Borax 1975. Being developed by Sandoz Agro.

TOXICITY: LD_{50} - 5,000 mg/kg. May cause eye and skin irritation.

FORMULATION: 65% WDG.

USES: Turf and ornamentals. Experimentally being tested on tree, nut and vine crops. Used outside the U.S. on alfalfa, cotton, soybeans, orchards, vineyards and other crops.

IMPORTANT WEEDS CONTROLLED: Barnyardgrass, foxtails, crabgrass, seedling johnsongrass, lambsquarters, pigweed, and many others.

RATES: Applied at .375-1.5 kg a.i./ha.

APPLICATION:

1. Preplant Incorporated—Incorporate into the soil 1-1.5 inches deep immediately after application.

2. Preemergence—Apply and immediately irrigate into the soil with an inch or more of sprinkler irrigation water. Apply in the fall or winter months. Used on turf prior to crabgrass emergence.

PRECAUTIONS: Agitate while applying. Do not use on bentgrass turf. Toxic to fish.

ADDITIONAL INFORMATION: Water solubility is .05 ppm. Gives season long weed control.

NAMES

ETHALFLURALIN, CURBIT, EDGE, SOMILAN, SONALAN, SONALEN

N-ethyl-N-(2-methyl-2-propenyl)-2,6 dinitro-4-(trifluoromethyl)benzenamine

TYPE: Ethalfluralin is a dinitroaniline compound used as a selective, soil-incorporated herbicide.

ORIGIN: 1973. Elanco Products, now DowElanco. Also marketed by Platte Chemical Co.

TOXICITY: LD_{50} - 10,000 mg/kg. May cause eye and skin irritation.

FORMULATION: 3 EC.

USES: Soybeans, dry beans, sunflowers, peanuts and cucurbits. Outside the U.S. it is also used on cotton and rape.

IMPORTANT WEEDS CONTROLLED: Barnyardgrass, crabgrass, foxtails, kochia, lambsquarter, nightshade, johnsongrass, panicum, chickweed, ground cherry, henbit, pigweed, purslane, and others.

RATES: Applied at .5-1.5 Ib. ai/A.

APPLICATION: Applied as a preplant soil incorporated herbicide to dry beans and soybeans. Incorporate into the top 2-3 inches of soil prior to planting within 2 days of application. Treated soil may be cultivated without reducing the activity. In some areas, this product can be surface applied as a preemergence herbicide. Rainfall or overhead irrigation (1/2 inch) is required within 2-5 days of application. On cucurbits used as a preemergence surface applied herbicide.

PRECAUTIONS: Do not soil incorporate prior to planting of cucurbits. Do not tank-mix with fertilizer when used on cucurbits. Do not use on cucurbits in association with plastic or hot cap cultural practices. Do not use the higher rates on garden beans, green beans, snap beans, or string beans. Certain crops should not be rotated within 8 months. Do not use on soils containing more than 10% organic matter.

ADDITIONAL INFORMATION: May be tank mixed with liquid fertilizer or other herbicides. May be impregnated on dry bulk fertilizer. Water solubility is .2 ppm.

NAMES

PENDIMETHALIN, GOGOSAN, HERBADOX, PRE-M, PROWL, STOMP, WAY-UP, SOVEREIGN, PENDULUM

$$CH_3 — CH_2 — CH — CH_2 — CH_3$$

(structure: 1-ethylpropyl group attached via NH to a benzene ring bearing two NO_2 groups at the 2,6 positions and two CH_3 groups at the 3,4 positions)

N-(1-ethylpropyl) 3,4-dimethyl-2,6-dinitrobenzenamine

TYPE: Pendimethalin is a dinitroaniline compound used as a preemergence and preplant herbicide.

ORIGIN: 1972. American Cyanamid Company.

TOXICITY: LD_{50} - 1250 mg/kg.

FORMULATIONS: 4EC. 60% DG. 3.3 EC. Formulated with other herbicides.

USES: Cotton, corn, garlic, grain sorghum, peanuts, non cropland, turf, ornamentals, Christmas trees, grapes, lima beans, snap beans, non-bearing fruil trees, rice, beans, peas, wheat, potatoes, and soybeans. Used on these and a number of other crops outside the U.S. Used to control suckers on tobacco outside the U.S.

IMPORTANT WEEDS CONTROLLED: Foxtails, barnyardgrass, panicum, crabgrass, pigweed, velvetleaf, lambsquarters, purslane, johnsongrass (from seed), and others.

RATES: Applied at .5-1.5 lb. ai/acre.

APPLICATION:Applied as a preplant incorporated herbicide or as a preemergence treatment depending upon the crop and usage. Rainfall or irrigation is required to move it into the soil when used preemergence. On turf apply prior to crabgrass germination.

PRECAUTIONS: Injury will occur to corn if used as a preplant, soil-incorporated treatment. Toxic to fish. Do not store below 40°F.

ADDITIONAL INFORMATION: May be tank mixed with other herbicides for broad spectrum control. Tolerant weeds include sicklepod, cocklebur, morningglory, jimson-weed, teaweed, ragweed, sesbania, nutgrass, and others. Wild cane and smartweed are semi-tolerant. Stronger on grasses than on broadleaf weeds. Inhibits germination and seedling development of susceptible plant species. Resists leaching in the soil. May be applied by air. Water solubility is .5 ppm.

NAMES

DIFLUFENICAN, COUGAR, QUARTZ, FENIKAN, LAZERIL, LUIZOR, FIRST, LAZERIL, IONIZ, BRODAL, JAVELIN

N-(2,4′-difluorophenyl)-2-(3-trifluoromethylphenoxy)
pyridine-3-carboxamide

TYPE: Diflufenican is a carboxamide compound used as a selective preemergence herbicide.

ORIGIN: May and Baker of England, 1979. Being marketed by Rhone Poulenc.

TOXICITY: LD_{50} - 2000 mg/kg.

FORMULATIONS: WP, aqueous suspensions and with other herbicides.

USES: Outside the U.S. on winter cereals, peas, lupines and sunflowers.

IMPORTANT WEEDS CONTROLLED: Pigweed, mustards, shepherd's purse, galinsoga, smartweed, purslane, teaweed, chickweed, and many others.

RATES: Applied at 125-250 g ai/ha.

APPLICATION: Applied as an overall spray to the soil surface either preemergence or early postemergence. The soil should be moist at the time of application. The soil should not be disturbed after application.

PRECAUTIONS: Not for sale or use in the U.S. Grasses are generally resistant to this product. Do not apply to soils over 10% organic matter. Do not soil incorporate. Do not apply to weeds after the 4 leaf stage. Some yellowing to barley has occurred, but it will outgrow it in 7-14 days.

ADDITIONAL INFORMATION: This is a residual contact herbicide translocated to a limited extent through the root system. May be mixed with other herbicides to improve the spectrum of control. Susceptible plants germinate, but show immediate chlorosis followed by a pink discoloration. Generally quick acting, but under cool growing conditions, complete control may take 6-8 weeks. Absorbed by the shoot of the weed as it comes in contact with the herbicide on the soil surface. The speed of the action is directly related to light intensity.

NAMES

FLUMIOXAZIN, S-53482

N-(7-fluoro-3,4-dihydro-3-oxo-4-prop-2-ynyl-2H-1,4-benzoxazin-6-yl) cyclohex-1-ene-1,2-dicarboximide

TYPE: Flumioxazin is a phthalimide compound used as a selective pre and postemergence herbicide.

ORIGIN: Sumitomo Chemical Co. of Japan 1989. Being developed in the U.S. by Valent.

TOXICITY: LD_{50} 5000 mg/kg. May cause eye irritation.

FORMULATION: 50% WDG.

USES: Experimentally being tested on soybeans, peanuts, rice, cereals, corn and other crops.

IMPORTANT WEEDS CONTROLLED: Velvetleaf, sicklepod, morningglory, smartweed, nightshade, cocklebur, and many other broadleaf weeds.

RATES: Applied at 50-100 g a.i./ha.

APPLICATION: Applied as both a preemergence and as a non selective postemergence treatment.

PRECAUTIONS: Used on an experimental basis only. Cotton is especially sensitive to postemergence applications of this material.

ADDITIONAL INFORMATION: Some grasses are controlled. Selectivity is not as great when used postemergence. Fast acting, since it is readily absorbed into the plant tissue.

NAMES

FLUPOXAM, MON 18500, KNW-739

1-[4-chloro-alpha-(2,2,3,3,3-pentafluoropropoxcy)-m-tolyl]-5-phenyl-1H-1,2,4-triazole-3-carboxamide

TYPE: Flupoxam is a triazole compound used as a selective pre and postemergence herbicide.

ORIGIN: Kureha Chemical Co. of Japan 1987. Being developed jointly with Monsanto.

TOXICITY: LD$_{50}$ 5000 mg/kg.

FORMULATION: 450 SC.

USES: Experimentally being tested on cereals.

IMPORTANT WEEDS CONTROLLED: Many annual broadleaf species.

RATES: Applied at 100-200 g a.i./ha.

APPLICATION: Applied as a pre or postemergence treatment.

PRECAUTION: Used on an experimental basis only. Grasses are not controlled.

ADDITIONAL INFORMATION: Being tested in combination with grass control herbicides. Gives contact and residual control. Activated by rainfall.

NAMES

BENSULIDE, **BENSUMEC, BETASAN, EXPORSAN, PREFAR, ROMPER**

0,0-bis (1-methylethyl) S-[2-[(phenylsulfonyl) amino] ethyl]
phosphoridithioate

TYPE: Bensulide is sulfonamide compound used as a preemergence, selective herbicide.

ORIGIN: 1962. Stauffer Chemical Company. ICI is the basic producer today. Being marketed also by Gowan Co.

TOXICITY: LD_{50} - 270 mg/kg.

FORMULATIONS: 4EC, 7% and 12.5% granules.

USES: Cotton, broccoli, Brussels sprouts, cabbage, carrots, grasses grown for seed, cauliflower, melons, onions, watermelons, lettuce, cucumbers, squash, ornamentals, turf, peppers, tomatoes, and dichondra. Used on rice in Japan.

IMPORTANT WEEDS CONTROLLED: Crabgrass, barnyardgrass, fall panicum, annual bluegrass, junglerice, foxtail, pigweed, goosegrass, purslane, and others.

RATES: Applied at 2-15 Ib active/A. Lower rates are used on cropland and the higher rates for turf weed control.

APPLICATION:

1. Turf—Apply at a uniform rate early in the spring or in the late fall to dichondra or established turf before the crabgrass geminates. Control will last through the entire growing season. The rate required is determined by lhe Iength of the growing season and the amount of water applied. Sprinkle for 10-15 minutes after treatment to move the material into the soil.

2. Cropland—Apply as a preplant, soil-incorporated treatment. Incorporate to a depth of 2-4 inches before planting. Apply to well-worked soil that is dry enough to permit incorporation. A preemergence, surface application can also be used if crops are to be irrigated up, either by sprinkler or furrow irrigation.

PRECAUTIONS: Apply to lawns only when they are well established. Temporary yellowing may occur to Bermuda turf. Do not reseed turf sooner than 4 months after treatment. A long-residual herbicide, so injury to crops may occur following a treated crop. Do not plant crops other than those on the label for a period of 18 months. Use on mineral soil only. Do not apply with liquid fertilizers.

ADDITIONAL INFORMATION: Dichondra is very tolerant to preplant, preemergence, and postemergence applications. A large number of ground covers and ornamentals have been treated and found to be tolerant. Must be applied before the weeds emerge. Six to eight months' control may be expected. Fairly resistant to leaching.

RELATED COMPOUNDS:
Diphenamid, Enide—Originally introduced by Elanco Products Co. and The Upjohn Co., it is now available in Europe through ICI as a preemergence herbicide for use on tomatoes, peppers, tobacco and other crops.

NAMES

NAPROPAMIDE, DEVRINOL, HURDLE, KUSALESS

$$CH_3 - CH - \overset{\overset{\displaystyle O}{\parallel}}{C} - N \overset{\diagup CH_2 - CH_3}{\diagdown CH_2 - CH_3}$$

N,N-diethyl-2-(1-naphthalenyloxy)propionamide

TYPE: Napropamide is a propionamide compound used as a selective, preemergence herbicide.

ORIGIN: 1969. Stauffer Chemical Company. ICI is the principle producer today.

TOXICITY: LD_{50} - 5000 mg/kg.

FORMULATIONS: 50% WP, 2EC, 50%DF. 2,5,and 10% granules.

IMPORTANT WEEDS CONTROLLED: Bromes, cheatgrass, foxtail, sprangletop, wild barley, cupgrass, annual bluegrass, crabgrass, barnyardgrass, prickly lettuce, filaree, groundsel, chickweed, fiddleneck, knotweed, lambsquarters, malva, pigweed, purslane, sowthistle, and others.

USES: Apples, almonds, apricots, artichoke, asparagus, avocados, blueberries, cherries, cole crops, dichronda, eggplant, figs, grapes, kiwi, mint, nectarines, olives, oranges, ornamentals, peaches, pears, peppers, persimmons, pistachios, plums, pomegranates, prunes, rhubarb, strawberries, tomatoes, pecans, rape, turf, and walnuts, in the U.S. Outside the U.S. It is used on these and other crops.

RATES: Applied at 2-6 kg a.i./ha.

APPLICATION:
Applied as a preplant incorporated or as a preemergence treatment. Apply to established turf prior to crabgrass germination. May be tank mixed with other herbicides.

PRECAUTIONS: Weeds not controlled include bindweed, annual morningglory, mustards, nightshade, sunflower, and turkey mullein. Do not use on soil over 10% in organic matter. Do not seed alfalfa, lettuce, small grains, sorghum, com, or sugar beets for 12 months after application. Toxic to fish.

ADDITIONAL INFORMATION: A long-lasting compound which will give season-long weed control. Use higher rates for perennial weed control. Water solubility is 73 ppm. Use on mineral soils only. Control weeds before they emerge, or apply with a contact herbicide. May be used on dichronda at seeding or on the established plants. May be used on container-grown ornamentals. May be applied through the irrigation water.

NAMES

DIMETHENAMID, **FRONTIER, SAN-582**

(*1RS*, a*RS*)-2-chloro-*N*-(2,4-dimethyl-3-thienyl)-*N*-(2-methoxy-1-methylethyl)-acetamide

TYPE: Dimethenamid is an acetamide compound used as a selective preemergence herbicide.

ORIGIN: Sandoz of Switzerland 1988.

TOXICITY: LD_{50} 1570 mg/hg. May cause slight eye and skin irritation.

FORMULATION: 7.5 EC.

USES: Experimentally being tested on corn, soybeans, sorghum, dry beans, peanuts, and other crops.

IMPORTANT WEED CONTROLLED: Crabgrass, barnyardgrass, panicum, foxtails, seedling johnsongrass, pigweed, spurge, purslane, nightshade, yellow nutsedge and others.

RATES: Applied at .85-1.7 kg a.i./ha. (.75-1.5 lbs a.i./A).

APPLICATION: Applied as a preplant, preplant incorporated or preemergence treatment. If rainfall does not occur in 7-10 days the material should be incorporated into the top 1-2 inches of soi. May be applied with liquid fertilizers.

PRECAUTIONS: Used on an experimental basis only.

ADDITIONAL INFORMATION: Controls weeds by reducing plant cell division and growth. They die before or soon after they emerge from the soil. Emerged weeds are not controlled. Does not persist in the soil.

NAMES

NAPROANILIDE, URIBEST

alpha-(2-napthyloxy)propionanilide

TYPE: Naproanilide is an anilide compound used as a postemergence, selective herbicide.

ORIGIN: 1969. Mitsui Toatsu Chemicals of Japan.

TOXICITY: LD_{50} - 15,000 mg/kg.

FORMULATIONS: 10% granules. Formulated with other herbicides in the granular form.

USES: Used in Japan on rice.

IMPORTANT WEEDS CONTROLLED: Bulrush, spikerush, Cyperus spp., water plantain, water starwort, and other broadleaf weeds and sedges.

RATES: Applied at 2-3 kg a.i./ha.

APPLICATION: Applied to weeds in the early growth stage either preemergence or postemergence before the 2.5 leaf stage. Applied after 3 days after transplantmg.

PRECAUTIONS: Not for sale or use in the U.S. No effect on barnyardgrass. Injury to rice sometimes occurs with high temperatures.

ADDITIONAL INFORMATION: Absorbed through both the stems and roots. Used in combination with other grass herbicides.

RELATED COMPOUNDS:
Monalide, Potablan—An anilide compound developed in 1964 by Sandoz and sold outside the U.S. for usage on carrots, celery and similar crops to control broadleaf weeds as a postemergence treatment.

Pentanochlor, Solan—A anilide compound introduced in 1960 by FMC and sold outside the U.S. by Atlas Interstates to use on carrots, parsley, celery, strawberries, tomatoes and ornamentals as a postemergence application.

NAMES

DIMETHACHLOR, **TERIDOX**

2,6-dimethyl-N-2′-methoxyethylchloroacetanilide

TYPE: Dimethachlor is an anilide compound used as a selective preemergence herbicide.

ORIGIN: 1973. CIBA-Geigy Limited.

TOXICITY: LD_{50} - 1600 mg/kg. May cause slight eye and skin irritation.

FORMULATION: 500EC.

USES: Rape-seed (winter varieties).

IMPORTANT WEEDS CONTROLLED: Blackgrass, foxtails, silky bentgrass, ryegrasses, annual bluegrass, shepherdspurse, fumitory, henbit, chamomile, chickweed, spurry, speedwell, vetches, lambsquarters, and others.

RATES: Applied at 1.25-2 kg ai/ha.

APPLICATION: Apply immediately after sowing before weeds or crop emerge.

PRECAUTIONS: Not for sale or use in the U.S.

ADDITIONAL INFORMATION: Taken up mainly through the shoots of germinating plants and seedlings. Weeds are mostly killed before emergence or sometimes shortly afterwards. Water solubilily: 2100 ppm. On light soils and under heavy rainfall conditions, stunting of rape may occur.

NAMES

CLOMEPROP, YUKAHOPE

(RS)-2-(2,4-dichloro-m-tolyloxy)propionanilide

TYPE: Clomeprop is an anilide compound used either as a selective preemergence or an early postemergence herbicide.

ORIGIN: 1980. Mitsubishi Petrochemical Co., Ltd.

TOXICITY: LD_{50} - 5,000 mg/kg

FORMULATION: Formulated wilh other herbicides in the granular form.

USES: Paddy rice

IMPORTANT WEEDS CONTROLLED: Bulrush, spikerush, sedges, Cyperus spp., water plantain, water starwort, and other annual broadleaf weeds.

RATES: Applied at 0.24-0.45 kg. ai/ha.

APPLICATION: Applied as a preemergence or an early postemergence, either before, or generally 3-7 days after transplanting.

PRECAUTION: Not for sale or use in the U.S.

ADDITIONAL INFORMATION: Symptoms are that of a phenoxy type herbicide like 2,4-D. Used in combination with other grass herbicides.

NAMES

PROPANIL, DIPRAM, ERBAN, GRASCIDE, HERBAX, PROPANEX, PROPANID, RISELECT, ROGUE, STAM, STAMPEDE, STREL, SUPERNOX, SURCOPUR, SYNPRAN, WHAM-EZ, PROSTAR, SORPUR, TURF EZ, DROPAVEN, RICELECT, CHEM-RICE, HERBAX

N-(3,4-dichlorophenyl) propanamide

TYPE: Propanil is an amide compound used as a selective, postemergence, contact herbicide.

ORIGIN: 1959. Rohm and Haas Company. Produced by numerous formulalors today.

TOXICITY: LD_{50}- 1384 mg/kg.

FORMULATIONS: 3EC, 4EC. Formulated with other herbicides.

USES: Rice, barley and wheat. Experimentally being used on turf.

IMPORTANT WEEDS CONTROLLED: Watergrass, pigeongrass, mustard, barnyardgrass, junglerice, nutsedge, Texas millet, crabgrass, paragrass, goosegrass, Brachiaria, jointed morningglory, pigweed, water plantain, and others. Tolerant weeds include sprangletop, red rice, lambsquarters, and kochia.

RATES: Applied at 1.12-6 Ib actual/A.

APPLICATION: Rice—Applied postemergence when barnyardgrass is in the 1-leaf to early-tiller stage, and before rice is in the late-tiller stage (45-60 days after planting). Barnyard grass is easiest to control in the 1-4-leaf stage. Do not apply to the second rice crop where double cropping is done. Treated fields may be reflooded 1-5 days after treatment to prevent reinfestation.

Wheat—Apply when the foxtails are in the 2-4-leaf stage. This is usually 10-17 days after crop emergence. Temporary yellowing may occur, but it will be outgrown.

PRECAUTIONS: Damage will occur if drift goes to nearby crops. A slight leaf burn to rice may result after treatment, but the rice plant will quickly outgrow it. Do not apply during extremely hot or cool weather. Do not use on rice at below 65°F or above 100°F temperature. Do not apply Sevin or phosphate insecticides to treated crop within 14 days before or after treatment. Do not use in combination with liquid fertilizer. Do not apply to rice later than 60 days after planting. Toxic to fish. Do not apply to wheat after the 5-leaf stage. Do not graze the treated fields. Do not apply to wheat grown on soils which had organic phosphate insecticides applied to them or to the previous crop. Do not apply to wheat when you expect daily temperatures to go below 50°F, or higher than 85°F. Do not apply to wheat growing under adverse conditions. Rainfall within 4 hours of application may reduce the degree of control.

ADDITIONAL INFORMATION: Not effective when applied preemergence. No residual effects. Soil type does not influence its action. Speed of kill is increased by an increase in temperature. Response declines as the temperatures go below 75°F. Apply in a medium-fine droplet spray, not as a coarse spray used on other materials. Nonvolatile. Most active on succulent, actively growing weeds. Most effective under conditions of high humidity. All leading commercial varieties of rice are extremely tolerant to this material. Temporary yellowing occurs on wheat, 2-5 days after application.

RELATED MIXTURES:

1. STAMPEDE CM—A combination of propanil and MCPA developed by Rohm and Haas to use on small grains.

NAME

HW-52

2,3-dichloro-4-ethoxymethoxy benzanilide

TYPE: HW-52 is an anilide compound used as a selective pre and early postemergence herbicide.

ORIGIN: Hodogaya Chemical Co. of Japan 1990.

TOXICITY: LD_{50} 5000 mg/kg. May cause slight eye irritation.

FORMULATION: 20% F, 50% WP, 3-7% granules.

USES: Experimentally being tested on rice.

IMPORTANT WEEDS CONTROLLED: Barnyardgrass.

RATES: Applied at 1-2 kg a.i./ha.

APPLICATION: Applied to both seeded and transplant rice grown under flooded conditions. Apply preemergence up to the 2 leaf stage of barnyardgrass.

PRECAUTIONS: Used on an experimental basis only. Avoid treatment to muddy water. Keep the field flooded for 2 weeks after treatment.

ADDITIONAL INFORMATION: Rice is extremely tolerant. Half life in the soil is 2-3 days so there should be no carry over problems.

NAMES

PROPACHLOR, ALBRASS, RAMROD, SATECID, FLASH, SENTINEL, PROLEX, SATECID, AMBER, SCANNER

2-chloro-N-(1-methylethyl)-N-phenylacetamide

TYPE: Propachlor is an acetamide compound being used as a selective, preemergence herbicide.

ORIGIN: 1964. Monsanto Company.

TOXICITY: LD_{50} - 550 mg/kg. Moderately irritating to the eyes.

FORMULATIONS: 20% granules, 4 EC. Formulated with other herbicides.

USES: Corn, cotton, sorghum, and soybeans. Used on these and other crops outside of the U.S.

IMPORTANT WEEDS CONTROLLED: Foxtails, barnyardgrass, sandbur, crabgrass, groundsel, annual ryegrass, goosegrass, pigweed, ragweed, lambsquarters, purslane, smartweed, wild buckwheat, carpetweed, Florida pusley, and others.

RATES: Applied at 4-6 Ib actual/A.

APPLICATION: Apply to the soil surface prior to weed and crop emergence. Apply only to a well-prepared seedbed. Rainfall or irrigation is then required to activate the chemical. One-third to three-fourths inch of rainfall is required. May be applied postemergence to corn before weeds reach the 2-leaf stage. On furrow irrigated corn, it may be incorporated into the soil, not more than 2 inches deep.

PRECAUTIONS: Johnsongrass, bindweed, Canada thistle, quackgrass, cocklebur, and bullnettle will not be controlled. If adequate rainfall does not occur, weed control will be lessened. Toxic to fish.

ADDITIONAL INFORMATION: One of the few herbicides that works in peat and muck soils. Eight weeks' control may be expected. Best results are obtained when moisture occurs within 10 days of application. Water solubility is 700 ppm. Works well under dry conditions. May be applied with liquid fertilizer or mixed with other herbicides.

NAMES

ALACHLOR, ALANEX, ALANOX, ALAZINE, LASSO, LAZO, PILLARZO, SATOCHLOR, STAKE, SADDLE, JUDGE, CHLORAL, CAVALIER, CROPSTAR, RENEUR, PARTNER, STALL, ADEOCHLOR, MICRO-TECH, CHIMICLOR, RALCHLOR

2-chloro-2',-6'-diethyl-N-(methoxymethyl) acetamide

TYPE: Alachlor is an acetamide compound used as a selective preemergence herbicide.

ORIGIN: 1967. Monsanto Company.

TOXICITY: LD_{50} - 930 mg/kg. Causes eye and skin irritation.

FORMULATIONS: 4EC, 15% granules, 4ME, (micro-encapsulated liquid.)

USES: Corn, cotton, dry beans, lima beans, peanuts, peas, sorghum, soybeans, woody ornamentals and sunflower. Used on these and other crops outside the U.S.

IMPORTANT WEEDS CONTROLLED: Purslane, goosegrass, carpetweed, Florida pusley, pigweed, barnyardgrass, crabgrass, foxtails, fall panicurn, witchgrass, lambsquarters, and yellow nutgrass.

RATES: Applied at 1.5 - 4 lb actual/A.

APPLICATION: Apply preemergence if rainfall is available to move the chemical into the soil. Otherwise, use a preplant application, incorporating into the soil 1/2-2 inches deep.

PRECAUTIONS: Susceptible crops include sugar beets and cucurbits. Weeds not controlled include bermudagrass, bullnettle, Canada thistle, established johnsongrass, bindweed, and quackgrass. Do not use on cotton, except in the areas designated on the label.

ADDITIONAL INFORMATION: Postemergence applications will give control, provided they are applied before grasses are in the 1-2-leaf stage. Requires more moisture for activation than does propachlor. Soybeans are extremely tolerant. Does not carry over in the soil. Best results are obtained when rainfall occurs within 10 days after application. Rainfall in the amount of .3-.6 inch is needed to activate the material. Water solubility is 242 ppm. Can be tank mixed with a number of other herbicides. Can be mixed with the liquid fertilizer. May be aerial applied. Persists in the soil for 6-10 weeks. May be impregnated on bulk dry fertilizer.

RELATED MIXTURES:

1. BRONCO—A combination of alachlor and glyphosate developed by Monsanto to be used preemergence on no-till soybeans and corn.

2. BULLET—A combination of alachlor and atrazine developed by Monsanto for use on corn.

3. FREEDOM—A combination of alachlor and trifluralin developed by Monsanto for use on soybeans as a preplant incorporated treatment.

4. LARIAT—A combination of alachlor and atrazine developed by Monsanto for use as a preemergence treatment on corn and sorghum.

NAMES

BUTACHLOR, BUTANEX, BUTANOX, LAMBAST, MACHETE, TRAPP, AIMCHLOR, TEER

2-chloro-2',6'-diethyl-N-(butoxymethyl) acetamide

TYPE: Butachlor is an acetamide compound used as a preemergence and preplant, selective herbicide.

ORIGIN: 1969. Monsanto Company.

TOXICITY: LD_{50} - 2000 mg/kg. Irritating to skin and eyes.

FORMULATIONS: 5EC, 5% granules.

IMPORTANT WEEDS CONTROLLED: Foxtails, crabgrass, goosegrass, barnyard grass, pigweed, lambsquarters, ragweed, purslane, nutgrass, cheatgrass, fall panicum, seedling johnsongrass, and others.

USES: Used on rice outside of the U.S . Experimentally being tested on cotton, peanuts, sugarbeets, rape and other crops.

RATES: Applied at 1-4.5 kg a.i./ha.

APPLICATION: Applied either preplant incorporated, or preemergence to seeded or transplant rice. Applied 3 -7 days after transplanting. Also applied postemergence 10-12 days after emergence.

PRECAUTIONS: Not for sale or use in the U.S. Toxic to fish. Do not treat if rain is expected within 6 hours.

ADDITIONAL INFORMATION: Solubility in water is 20 ppm. Soil incorporation is suggested for drought conditions. Depth of incorporation should not exceed 2 inches.

58

Shows effective postemergence activity against young grasses up to the 2-leaf stage. Persists in the soil for not over 10 weeks. Can be tank mixed with other herbicides.

NAMES

METOLACHLOR, **DUELOR, MEDAL, FALCON, PENNANT**

2-chloro-N-(2-ethyl-6-methylphenyl)-N-(2-methoxy-1-methylethyl)acetamide

TYPE: Metolachlor is an acetamide compound used as a selective, preemergence herbicIde.

ORIGIN: 1974. CIBA-Geigy Chemical Company.

TOXICITY: LD_{50} - 2780 mg/kg. May cause eye and skin irritation.

FORMULATIONS: 8EC, 5G, 500EC, 720EC. 7.8EC. Formulated with other herbicides.

USES: Corn, cotton, ornamentals, pod & seed vegetables, pecans, cabbage, potatoes, peanuts, safflower, sorghum, soybeans. Also used in non-crop weed control. Used outside the U.S. on these plus, sugarcane, sugar beets, sunflowers, and others.

IMPORTANT WEEDS CONTROLLED: Barnyardgrass, chickweed, crabgrass, foxtails, galinsoga, goosegrass, millets, nightshade, panicum, pigweed, purslane, signalgrass, smartweed, red rice, yellow nutsedge, and others.

RATES: Applied at 1-4 Ib actual/A.

APPLICATION: Applied either preemergence or preplant incorporated. If rainfall does not occur within 10 days, incorporate lightly. Under furrow irrigation, it can be preplant incorporated in the top 2 inches of soil. May be applied 45 days prior to planting. Used on established ornamentals at least 5 days after transplanting.

PRECAUTIONS: Toxic to fish. Do not use on sweet potatoes or yams. Injury may occur under abnormally high soil moisture conditions. Do not apply to the wet foliage of ornamentals. Do not use on grain sorghum unless it has been treated with Concep seed safener. Do not use on English peas grown in the Northeastern U.S.

ADDITIONAL INFORMATION: Water solubility is 530 ppm. May be applied in combination with many other herbicides depending upon the crop. May be applied by air or through center pivot irrigation equipment. Taken up through the shoots of germinating plants and seedlings. May be applied with liquid fertilizers or impregnated on dry fertilizer.

RELATED MIXTURES:

1. BICEP—A 1.25:1 ratio of metolachlor and atrazine sold by Ciba Geigy in the U.S. for use on corn and sorghum.

2. CYCLE—A 1:1 ratio of metolachlor and cyanazine sold in the U.S. by Ciba Geigy for use on corn and sorghum.

3. TURBO EC—A metolachlor-metribuzin combination marketed by Miles Inc. to use as a preplant or preemergence herbicide on soybeans and potatoes.

4. DERBY—A metolachlor-simazine combination marketed by Ciba Geigy for usage on ornamentals.

NAMES

BROMOBUTIDE, SUMIHERB, LYTON

(R,S)-2-bromo-N-(alpha, alpha-dimethylbenzyl)-3,3-dimethylbutyamide

TYPE: Bromobutide is a amide compound used as a selective pre and postemergence herbicide.

ORIGIN: 1980. Sumitomo Chemical Co. of Japan.

TOXICITY: LD_{50} - 5,000 mg/kg. May cause slight eye irritation.

FORMULATIONS: 6% granules. Formulated with other herbicides for broadleaf control.

USES: Outside the U.S. on rice.

IMPORTANT WEEDS CONTROLLED: Barnyardgrass, Monochoria, sedges and others.

RATES: Used at 30-40 kg/ha (6% granules).

APPLICATION: Applied preemergence or early postemergence 3-8 days after transplanting. Barnyardgrass should be less than the 1.5 leaf stage. Can be used on both transplant and direct seeded rice.

PRECAUTIONS: Not for sale or use in the U.S. Rice injury may occur on sandy fields or on rice that has been shallowly transplanted.

ADDITIONAL INFORMATION: Effective on grasses and sedges. Does not inhibit the germination of weeds, but inhibits their growth after emergence. Gives at least 6 weeks control.

NAMES

PRETILACHLOR, **RIFIT, SOLNET, SOFIT**

2-chloro-2',6'-diethyl-N-(2-propoxyethyl)-acetanilide

TYPE: Pretilachlor is an acetanilide compound used as a selective preemergence herbicide.

ORIGIN: 1982. CIBA-Geigy.

TOXICITY: LD_{50} - 6099 mg/kg. May cause some skin and eye irritation.

FORMULATIONS: EC 500, 2% granules. When formulated with the safener fencloran it is trade named SOFIT.

USES: Outside the U.S. on wet sown rice and on transplanted rice. It can be used on wet sown rice only when mixed with the safener.

IMPORTANT WEEDS CONTROLLED: Cyperus Spp., barnyardgrass, sprangletop, Scripus Spp., and others.

RATES: Applied at 125-1000 g ai/ha.

APPLICATION: Applied either pretransplanting or sowing or between the time of transplanting or sowing and weed emergence.

PRECAUTIONS: Not for sale or use in the U.S. Use only in combination with the safener on direct seeded rice. Not suitable for use on upland rice.

ADDITIONAL INFORMATION: Residual control should last for 30-50 days. The full herbicidal activity of this material is only obtained in saturated soil under permanent flooding.

NAMES

ACETOCHLOR, ACENIT, HARNESS, RELAY, WINNER, GUARDIAN, SURPASS

2-chloro-N-(ethoxymethyl)-N-(2-ethyl-6-methylphenyl) acetamide

TYPE: Acetochlor is an acetamide compound used as a selective, preemergence herbicide.

ORIGIN: 1980. Monsanto Company. Being developed by ICI.

TOXICITY: LD_{50} - 1426 mg/kg. May cause eye and skin irritation.

FORMULATION: 6.4 EC. Sometimes formulated with a safener. (dichlormid)

USES: Experimentally on corn, soybeans, peanuts, and sorghum in the U.S. Outside the U.S. it is used on soybeans and other crops.

IMPORTANT WEEDS CONTROLLED: Most annual grasses, yellow nutsedge, ragweed, pigweed and others.

RATES: Applied at 1-3 pints (formulation) per acre.

APPLICATION: Can be applied preemergence, preplant incorporated or early postemergence.

PRECAUTIONS: Do not use in mild steel spray tanks or use PVC or rubber for hoses or pipes. Used in the U.S. on an experimental basis only.

ADDITIONAL INFORMATION: Absorbed mainly by germinating plant shoots. Active on heavy or high organic matter soils. Herbicidal effectiveness is 8-12 weeks. May be mixed with other herbicides or applied in liquid fertilizer. May be applied by air. Usually .3-.6 inch of rainfall will activate the product if it occurs within 7-10 days. Water solubility is 233 ppm.

NAMES

BUTAM, TEBUTAM, COMODOR

N-benzyl-N-isopropyl trimethylacetamide

TYPE: Butam is an acetamide compound used as a selective, preemergence herbicide.

ORIGIN: 1981. Dr. R. Maag of Switzerland. (Now Ciba Geigy).

TOXICITY: LD_{50} - 6000 mg/kg. May cause eye and skin irritation.

FORMULATIONS: 720 g ai/L EC. Formulated with other herbicides.

USES: Oulside the U.S. on rape, soybeans, peanuts, tobacco, tomatoes, cole crops, sunflowers, and other vegetable crops.

IMPORTANT WEEDS CONTROLLED: Wild oats, blackgrass, crabgrass, barnyardgrass, ryegrass, poa annua, foxtails, volunteer cereals, pigweed, lambsquarters, smartweed, chickweed, and others.

RATES: Applied at 2.8-3.6 kg ai/ha.

APPLICATION: Applied either preplant incorporated or as a preemergence herbicide. Rainfall or irrigation after application improves the performance.

PRECAUTIONS: Not for sale or use in the U.S. Susceptible crops include cereals, carrots, corn, flax, onions, garlic, lettuce, melons, sorghum, spinach, and sugar beets.

ADDITIONAL INFORMATION: Inhibits germination and shows activity through the root system. Sometimes affected plants will emerge, but they are severely stunted. Tolerant weeds include quackgrass, nutsedge, shepherd's purse, bedstraw, wild radish, and others. Can be mixed with other herbicides. Short-lived in the soil so following crops are not effected.

NAMES

METAZACHLOR, BUTISAN S, PREE, TRACK

N-(2,6-dimethylphenyl)-N-(1 -pyrazolymethyl)-chloroacetamide

TYPE: Metazachlor is an acetanilide compound used as a selective preemergence herbicide.

ORIGIN: 1976. BASF of Germany.

TOXICITY: LD_{50} - 2150 mg/kg.

FORMULATIONS: 500 EC.

USES: Outside the U.S. on cotton, cole crops, corn, ornamentals, kale, rape, potatoes, peanuts, soybeans, tobacco and others.

IMPORTANT WEEDS CONTROLLED: Pigweed, lambsquarters, purslane, yellow nutsedge, galinsoga, bedstraw, chickweed, shepherd's purse, mayweed, pennycress, speedwell, blackgrass, crabgrass, barnyardgrass, goosegrass, foxtails, ryegrass, annual bluegrass, sandbur, jungle rice, and others.

RATES: Applied at 1-1.5 kg a.i./ha.

APPLICATION: Used as a preemergence herbicide prior to weed emergence. Rainfall before application and high soil moisture at time of application increases the activity. May be applied postemergence to rape up to the 4th leaf stage.

PRECAUTIONS: Not for sale or use in the U.S. Toxic to fish.

ADDITIONAL INFORMATION: On slight soils, heavy rainfall after application may cause some crop injury. To get maximum weed control, metazachlor should not be applied on a rough or cloddy seedbed. Water solubility is 500 ppm. May be mixed with other herbicides.

CARBAMATES

NAMES

CHLOROPROPHAM, CIPC,
CHLORO IPC, FURLOE, MIRVALE

1-methylethyl 3-chlorophenylcarbamate

TYPE: CIPC is a selective carbamate herbicide, applied preplant, preemergence, and early postemergence.

ORIGIN: 1950. PPG Industries, Inc. Being sold in Europe by Pan Britannica and others.

TOXICITY: LD_{50} - 3800 mg/kg.

FORMULATIONS: 3EC, 4EC. Formulated with other herbicides.

USES: Outside the U.S. on alfalfa, beans, blueberries, caneberries, carrots, peas, clover, cowpeas, cranberries, garlic, lettuce, onions, safflower, soybeans, spinach, perennial grass, peppers, potatoes, rice, sugar beets, tomatoes, and ornamentals. Also use as a growth regulator on potatoes to prevent sprouting.

IMPORTANT WEEDS CONTROLLED: Crabgrass, foxtails, barnyardgrass, ryegrass, purslane, dodder, smartweed, knotweed, bluegrass, bromes, chickweed, wild oats, and many others.

RATES: Applied at 2-8 Ib actual/A.

APPLICATION: Can be applied preplant, preemergence, or postemergence, depending upon the crop, weed species, soil type, and environmental conditions. Moisture is necessary in the soil to germinate the weed seeds, activate the chemical, and move it into the soil. May be applied with the irrigation water to some crops.

PRECAUTIONS: No longer used as a herbicide in the U.S. Do not apply to moist or dew-covered foliage. Do not use the herbicide formulation for potato sprout inhibition.

ADDITIONAL INFORMATION: Chickweed is controlled at almost any stage of growth. Does not effect germination, but inhibits the growth of the primary root. Residual life of 1-6 months, depending upon the temperature. Lost from the soil rapidly above 85°F and slowly below 50°F. The higher the exchange capacity of the soil, the higher should be the rate of application. Strongly absorbed by organic matter. Resistant to leaching. Most active in light, sandy soils. More active on grasses than on broadleaves.

RELATED COMPOUNDS:
Propham, IPC, Ban Hoe—An analog of CIPC developed by PPG Industries in 1945. It is still used to a limited extent in some countries as a preemergence herbicide on peas, sugarbeets, lettuce and other crops.

NAMES

EPTC, EPTAM, ERADICANE, CAPSOLANE, NIPTON, ALIROX

$$CH_3-CH_2-S-\overset{\overset{\displaystyle O}{\|}}{C}-N\overset{\displaystyle CH_2-CH_2-CH_3}{\underset{\displaystyle CH_2-CH_2-CH_3}{}}$$

S-ethyl dipropylcarbamothioate

TYPE: EPTC is a selective thiocarbamate herbicide applied preplant prior to weed germination.

ORIGIN: 1959. Stauffer Chemical Company. Being sold by ICI Ag Products.

TOXICITY: LD_{50} -1367 mg/kg. May cause skin and eye irritation.

FORMULATIONS: 7EC. 2,3,5 and 10% granules, 6 EC, 6.7 EC. Formulated with a safener for use on corn.

USES: Alfalfa, almonds, castorbeans, citrus, clovers, corn, cotton, dry beans, flax, grapes, lespedeza, ornamentals, peas, pine nurseries, potatoes, safflower, snap beans, sugar beets, sunflowers, sweel potatoes, table beets, tomatoes, trefoil, and walnuts. Used on these and other crops outside the U.S.

IMPORTANT WEEDS CONTROLLED: Nutgrass, johnsongrass (from seed), quackgrass, foxtails, sandbur, wild oats, barnyardgrass, chickweed, nightshade, lambsquarters, pigweed, henbit, purslane, and a number of other annual grass and broadleaf weeds.

RATES: Applied al 2-7.5 Ib active/A.

APPLICATION: Incorporate into the soil the same day of application. Applied preemergence or preplant, either broadcast or in bands. Incorporate with disc harrows or hooded power-driven rotary tillers 2-4 inches deep. If incorporated with a disc harrow, cross disc (twice at right angles) at a 4-6 inch depth . Existing perennial weeds should be turned under and chopped up thoroughly before treatment for satisfactory control . Also may be metered into the irrigation water on sugar beets, alfalfa, clover, almonds, walnuts, and grapes. Also applied in some areas by the subsurface injeclion technique developed for the area.

PRECAUTIONS: Readily lost from the soil by volatilization when the soil surface is wet at the time of application. Do not use on blackeyed beans, soybeans, lima beans, and other flat-podded beans. Corn is injured if planted deeper than 2 inches. Do not move untreated soil into the incorporated area.

ADDITIONAL INFORMATION: Not effective as a contact herbicide. More effective on grasses than broadleaves. Decomposes in 4-6 weeks in warm, moist soils. Use higher rates on peat or muck soils. Established plants tolerate over-the-top, spray application. Non-corrosive to metal or rubber. Applied in the irrigation water to a number of crops. May be applied preplant, soil-incorporated in the fall to sugar beets grown in some areas prior to the time the ground freezes. May be applied through sprinkler irrigation systems. May be used with liquid fertilizers or incorporated on dry bulk fertilizers.

NAMES

PEBULATE, PEBC, TILLAM

S-propyl butylethylcarbamothioate

TYPE: Pebulate is a selective thiocarbamate herbicide applied preplant or postplant and soil incorporated.

ORIGIN: 1959. Stauffer Chemical Company. ICI Ag Products is the basic producer today.

TOXICITY: LD_{50} - 1120 mg/kg.

FORMULATION: 6EC.

USES: Tomatoes, sugar beets, and tobacco.

IMPORTANT WEEDS CONTROLLED: Barnyardgrass, foxtail, crabgrass, nutgrass, wild oats, pigweed, purslane, lambsquarters, and others.

RATES: Applied at 4-6 Ib active/A.

APPLICATION: Applied preplant. Must be incorporated to a depth of 2-4 inches immediately after application. Incorporation can be with a hooded power-driven rotary tiller or with a disc harrow. Using a disc harrow, incorporate at a 5-6 inch depth by cross discing twice at right angles. Can be seeded immediately after treatment. For existing nutgrass control, turn under and chop up weeds thoroughly before application.

ADDITIONAL INFORMATION: Closely related to EPTC. Most effective on grasses. Broadleaves are killed if application is made when conditions are favorable for germination. No contact activity. If tomatoes are grown under hot caps, they should be vented to prevent injury. Not as effective on nutgrass as EPTC. Activity should last for 6-8 weeks. May be used with liquid fertilizers or impregnated on dry bulk fertilizers.

NAMES

CYCLOATE, RO-NEET, SABET

S-ethyl cyclohexylethylcarbamothioate

TYPE: Cycloate is a thiocarbamate compound, effective as a selective, soil incorporated herbicide.

ORIGIN: 1963. Stauffer Chemical Company. ICI Ag Products markets today.

TOXICITY: LD$_{50}$ - 2710 mg/kg.

FORMULATIONS: 6EC.

USES: Sugar beets, table beets, and spinach.

IMPORTANT WEEDS CONTROLLED: Ryegrass, nutgrass, barnyardgrass, crab-grass, henbit, annual bluegrass, foxtails, wild oats, nightshade, lambsquarters, pigweed, purslane, shepherd's purse, volunteer barley, slinging nettle, and others.

RATES: Applied al 3-4 Ib active/A.

APPLICATION: Apply to a well-prepared seedbed. Incorporate immediately into the soil to a depth of 2-3 inches. Plant immediately.

PRECAUTIONS: A volatile material, so incorporation must be done immediately after application. Do not band apply on rocky soils since thorough incorporation is impossible. Do not use a drag behind the planter as it may concentrate the chemical over the seed row. Use on mineral soils only. Do not combine with other pesticides. Do not apply before a preirrigation.

ADDITIONAL INFORMATION: Planting should be done immediately after incorporation. Apply only to soil dry enough to insure thorough mixing. Existing nutgrass can be controlled by turning under and chopping up weeds thoroughly before application. An analog of EPTC. Six to twelve weeks' control may be expected. May be applied in liquid fertilizer or impregnated on dry bulk fertilizer.

NAMES

BUTYLATE, ANELDA, SUTAN +

$$CH_3-CH_2-S-\overset{\displaystyle O}{\overset{\displaystyle \|}{C}}-N\begin{cases} CH_2-CH\begin{cases} CH_3 \\ CH_3 \end{cases} \\ CH_2-CH\begin{cases} CH_3 \\ CH_3 \end{cases} \end{cases}$$

S-ethyl bis (2-methylpropyl) carbamothioate

TYPE: Butylate is a thiocarbamate compound used as a selective preplant soil incorporated herbicide.

ORIGIN: 1959. Stauffer Chemical Company. ICI Ag Products markets it today.

TOXICITY: LD_{50} - 3500 mg/kg.

FORMULATIONS: 6.7 EC. Formulated with other herbicides. Formulated with a safener for use on corn.

USES: Corn.

IMPORTANT WEEDS CONTROLLED: Foxtails, johnsongrass (from seed), barnyard grass, wild cane, Texas panicum, goosegrass, nutgrass, watergrass, crabgrass, and others.

RATES: Applied at 3-4 Ib active/A.

APPLICATION: Incorporate within 4 hours into the soil with a power-driven rotary hoe or disc. Apply to a dry surface only. Incorporate 3-6 inches deep with power-driven tiller or disc and follow with a spike-toothed harrow. Disc in 2 directions. Existing stands of nutgrass should be turned under and chopped up prior to treatment. Subsurface injection methods of application are recommended in some areas. May be mixed with other herbicides.

PRECAUTIONS: Do not apply before a preirrigation. Do not use on hybrid corn grown for seed. Plant corn at least 2 inches deep. Recommended on mineral soils only, containing less than 10% organic matter.

ADDITIONAL INFORMATION: Closely related to EPTC. Six weeks' control can be expected, and there are no soil carryover problems affecting the following crop. No contact activity. May be applied in liquid fertilizer or impregnated on dry bulk fertilizer.

RELATED MIXTURES:

1. SUTAZINE—A combination of butylate and atrazine developed by ICl for use on corn. Formulated as a 6 EC and as a 18-6 granular.

NAMES

VERNOLATE, REWARD, VERNAM

$$CH_3 — CH_2 — CH_2 — S — \overset{\overset{O}{\|}}{C} — N \overset{\diagup CH_2 — CH_2 — CH_3}{\diagdown CH_2 — CH_2 — CH_3}$$

S-propyl dipropylcarbamothioate

TYPE: Vernolate is a thiocarbamate compound used as a selective, soil-incorporated herbicide.

ORIGIN: 1954. Stauffer Chemical Company. Being sold by Drexel Chemical Co.

TOXICITY: LD_{50} -1500 mg/kg.

FORMULATIONS: 7EC.

USES: Peanuts and soybeans.

IMPORTANT WEEDS CONTROLLED: Crabgrass, barnyardgrass, carpetweed, purslane, sicklepod, velvetleaf, foxtails, nutgrass, johnsongrass seedlings, goosegrass, pigweed, lambsquarters, morningglory, and others.

RATES: Apply at 2-3 Ib actual/A.

APPLICATION: Applied as a preplant soil incorporated treatment. Incorporate into the soil immediately.

PRECAUTIONS: Incorporation immediately is extremely necessary. Do not move untreated soil onto the treated band. Use only on mineral soils. Readily lost by volatilization if the soil surface is wet at the time of application. Do not tank mix with insecticides or fungicides.

ADDITIONAL INFORMATION: Closely related to EPTC, but recommended crops show a greater degree of tolerance while still maintaining adequate weed control. No soil residue problem, since it dissipates rapidly. May be combined with certain other herbicides. May be applied through sprinkler irrigation systems. May be mixed with liquid fertilizers or impregnated on dry bulk fertilizers.

NAMES

PYRIBUTICARB, TSH-888, EIGEN, TOYOCARB

0-3-tert-butylphenyl-6-methoxy-2-pyridyl (methyl) thiocarbamate

TYPE: Pyributicarb is a thiocarbamate compound used as a selective preemergence and postemergence herbicide.

ORIGIN: Tosoh Corp. of Japan 1983.

TOXICITY: LD_{50} 5000 mg/kg. May cause skin irritation.

FORMULATIONS: Sold formulated with other herbicides, such as bromobutide and benzofenap. 47% WP.

IMPORTANT WEEDS CONTROLLED: Echinochloa spp., Cyperus spp., Monochoria, Setaria spp. and others.

USES: Outside the U.S. on rice. Under development for use on turf.

RATES: Applied at 1 kg a.i./ha.

APPLICATION: Applied either preemergence or early postemergence when the weeds are not past the 2 leaf stage.

PRECAUTION: Not for sale or use in the U.S.

ADDITIONAL INFORMATION: Most effective against grasses. Mostly formulated with other herbicides to increase the spectrum of control. Long lasting. Being developed as a turf herbicide.

NAMES

ESPROCARB, ICIA-2957

S-benzyl 1,2-dimethylpropyl (ethyl) thiocarbamate

TYPE: Esprocarb is a thiocarbamate compound used as a pre and postemergence herbicide.

ORIGIN: Stauffer Chemical Co. 1969. Being developed by ICI.

TOXICITY: LD_{50} 4600 mg/kg.

FORMULATION: 5% granules. May be combined with bensulfuron-methyl.

USES: Used in Japan on rice.

IMPORTANT WEEDS CONTROLLED: Brachiaria, nutsedge, barnyardgrass, sprangletop and other weeds.

RATES: Applied at 1-2 kg a.i./ha.

APPLICATION: Applied either preemergence or early postemergence. Barnyardgrass will be controlled up to the 2.5 leaf stage. Apply within 5-20 days after transplanting.

PRECAUTION: Not for sale or use in the U.S. Toxic to fish.

ADDITIONAL INFORMATION: Mostly sold in combination with bensulfuron-methyl under the tradename Fujigrass. Water solubility is 4.9 ppm.

NAMES

CARBETAMIDE, CARBETAMEX, LEGURAME

N-ethyl-2[[phenylamino) carbonyl]oxy] proponamide

TYPE: Carbetamide is a carbamate compound used as a selective, pre and postemergence herbicide.

ORIGIN: 1960. Rhone Poulenc.

TOXICITY: LD_{50} - 11,000 mg/kg.

FORMULATIONS: 300 g/L EC, 70% WP. Formulated with other herbicides.

USES: Outside the U.S. on alfalfa, rape, lentils, chicory, strawberries, sugarbeets, onions, sunflowers, lettuce, vines, clover, orchards, and nurseries.

IMPORTANT WEEDS CONTROLLED: Annual grasses and quackgrass.

RATES: Applied at 2 kg a.i./ha.

APPLICATION: Apply to the soil surface during the winter months on established fields or on fall seedlings. Apply also in the spring on sprouting grasses, if they are not past the first trifoliate leaf. Water must take it into the soil.

PRECAUTIONS: Not sold in the U.S. Most effective in cool weather. Slow acting.

ADDITIONAL INFORMATION: Due to its solubility it does not need soil incorporation. Absorbed by the roots and somewhat by the leaves of grass plants. Activity lasts 2-3 months in the winter and about I month in the spring. Effective on only a few summer grasses.

NAMES

PHENMEDIPHAM, BETANAL, BETOSIP, GOLIATH, GUSTO, KEMIFAM, PISTOL, SPIN-AID, BETAFLOW, BEETUP, PROTRUM, BEETOMAX, BETALION, SUPLEX, BETA, VANGARD, AIMSAN, MEDIFENE

3-methoxycarbonylaminophenyl-3-methylphenylcarbamate

TYPE: Phenmedipham is a carbamate compound used as a selective, postemergence herbicide.

ORIGIN: 1968. Schering AG of Germany. Developed in the U.S. by Nor-Am Ag Products.

TOXICITY: LD_{50}- 8000 mg/kg.

FORMULATION: 1.3 EC. Also sold mixed with desmedipham in the U.S.

USES: Sugar beets, spinach, table beets and sunflowers in the U.S. Sold outside the U.S. on these plus sunflowers, strawberries and other crops.

IMPORTANT WEEDS CONTROLLED: Lambsquarters, mustard, pigweed, ragweed, kochia, chickweed, shepherd's purse, London rocket, wild buckwheat, nightshade, and others.

RATES: Applied at .75-1.5 lb a.i./A. Use higher rate on less-susceptible weeds.

APPLICATION: Apply as a postemergence application when the weeds are in the cotyledon to the 2-true leaf stage. Best results are obtained when the weeds are growing actively. Spray kochia when it is less than 1 inch in diameter.

PRECAUTIONS: Do not let the spray solution stand more than 4 hours in the spray tank. Do not apply when dew is present. Rainfall within 6 hours of application may reduce the effectiveness. Do not apply under extremely hot or windy conditions. Avoid drift. Do not apply with a wetting agent. May cause temporary growth retardation and/or chlorosis or tip burn to sugar beets. Toxic to fish. Do not apply to spinach when temperatures exceed 75°F.

ADDITIONAL INFORMATION: May be mixed with other pesticides. Absorbed through the leaves. Weeds are the most susceptible from the cotyledon to the 2-leaf stage. Soil breakdown occurs quite rapidly so there is no danger to following crops. Very little preemergence activity. Water solubility is less than 1 ppm. Effects will not be seen for 4-8 days after spraying.

NAMES

DESMEDIPHAM, BETANEX, BETANAL-AM

Ethyl [3-[[(phenylamino) carbonyl] oxy] phenyl] carbamate

TYPE: Desmedipham is a carbamate compound used as a selective, postemergence herbicide.

ORIGIN: 1970. Schering AG of Germany. Developed in the U.S. by Nor-Am.

TOXICITY: LD_{50} - 10,250 mg/kg.

FORMULATION: 1.3EC. Also sold mixed with phenmediphan.

USES: Sugar beets. Used outside the U.S. on these and other crops.

IMPORTANT WEEDS CONTROLLED: Pigweed, nightshade, sowthistle, lambsquarters, ragweed, purslane, mustard, shepard's purse and other broadleaf weeds.

RATES: Applied at 800-1000 g a.i./ha.

APPLICATION: Applied postemergence when the weeds are in the colyledon to the 2-leaf stage. Herbicidal activity becomes evident 4-8 days after treatment. Beets should have unfolded 3-4 true Ieaves prior to application. This is usually prior to thinning.

PRECAUTIONS: Toxic to fish. May cause injury if beets are under stress at the time of application. Do not apply to beets before they reach the 2-leaf stage. Do not use with additional surfactant. Rainfall within 6 hours of application may reduce the control. Avoid drift.

ADDITIONAL INFORMATION: Often tank mixed with phenmediphan at a 1:1 ratio to control weeds. More effective on pigweed species than phenmediphan.

RELATED MIXTURES:

1. BETAMIX --—A 50:50 combination of desmediphan and phenmediphan developed by Nor-Am for use on sugar beets.

NAMES

ORBENCARB, LANRAY

S-[2-(chlorophenyl) methyl] diethylcarbamothioate

TYPE: Orbencard is a carbamate compound used as a selective preemergence herbicide.

ORIGIN: Kumiai Chemical Ind. Co. of Japan. 1975.

TOXICITY: LD_{50} - 800 mg/kg.

FORMULATION: 50 EC. Also formulated with other herbicides.

USES: Used outside the U.S. on cereals and turf.

IMPORTANT WEEDS CONTROLLED: Crabgrass, nutsedge, barnyard grass, annual bluegrass, purslane, foxtails, goosegrass, bermudagrass and many others. More effective on grasses than broadleaves.

RATES: Applied at 4-5 kg a.i./ha.

APPLICATION: Applied as a preemergence herbicide.

PRECAUTIONS: Not for sale or use in the U.S. Somewhat toxic to fish.

ADDITIONAL INFORMATION: May be mixed with other herbicides to give better broadleaf control. Prevents the elongation of leaves after weed germination. Does not inhibit the roots.

NAMES

TRIALLATE, AVADEX-BW, FAR-GO

```
CH3
  \
   CH      O              Cl   Cl
  /         ||            |    |
CH3          N—C—S—CH2 —C ═ C
CH3        /                  |
  \       /                   Cl
   CH
  /
CH3
```

S-(2,3,3-trichloro-2-propenyl) bis(1-methylethyl) carbamothioate

TYPE: Triallate is a selective carbamate compound used as a preemergence herbicide.

ORIGIN: 1959. Monsanto Company.

TOXICITY: LD_{50} - 1100 mg/kg.

FORMULATIONS: 4EC, 10% granules.

USES: Barley, spring wheat, durum wheat, peas, chickpeas, lentils and winter wheat.

IMPORTANT WEEDS CONTROLLED: Wild oats.

RATES: Applied at 1-1.5 Ib actual/A.

APPLICATION: Apply to a well-worked seed bed and incorporate without delay to a depth of 2 inches. Seed should be placed 1/2-1 inch below the treated area. Applied preplant or preemergence. May be applied with other herbicides.

PRECAUTIONS: Crop thinning may occur, especially under high rainfall conditions following application. Do not apply to fields left in a ridged condition. Poor control may result if applied to a wet, cloddy, or rough soil. Tame oats should not be planted within twelve months of the date of application. Weed control may not be as effective under drought conditions. Leaf crinkling and delayed maturity may occur when applied to dry peas. Avoid drift.

ADDITIONAL INFORMATION: Applied either before or after seeding of barley and peas. However, on wheat, it should be applied after planting only. Peas may be planted up to three weeks after the application. Controls weeds only before they germinate. Do not incorporate deeper than two inches as dilution may occur. Under dry conditions, wild oats may reach the 2-4- leaf stage before being controlled. Shallow cultivation does not reduce the effectiveness. Fall applications may be made 2-3 weeks prior to soil freeze-up. May be applied by air. May be mixed with liquid fertilizers or impregnated on dry bulk fertilizers.

RELATED MIXTURES:

1. BUCKLE—A combination of triallate and trifluralin, marketed by Monsanto in the U.S. on wheat, barley and field peas as a preplant treatment.

NAMES

ASULAM, ASULOX, ASILAN

Methyl [(4-aminophenyl) sulfonyl) carbamate

TYPE: Asulam is a carbamate compound used as a selective, postemergence, systemic herbicide.

ORIGIN: 1968 May & Baker Co., of England. Being marketed by Rhone Poulenc, Inc.

TOXICITY: LD_{50} - 2000 mg/kg.

FORMULATION: 3.34 AS (sodium salt).

IMPORTANT WEEDS CONTROLLED: Johnsongrass, bracken, barnyardgrass, goosegrass, panicums, alexandergrass, dallisgrass, vaseygrass, crabgrass, foxtails, wild oats, Rumex species, horsetail, bracken fern, and others.

USES: Sugarcane, reforestation areas, turf, ornamentals, Christmas tree plantings, and non-crop areas. Used on these and pastures, bananas, rubber, flax and other crops outside the U.S.

RATES: Applied at 2-7 Ib active/A.

APPLICATION: Applied postemergence, when the weeds are growing vigorously. It can be taken up from both the roots and the leaves and translocated to other parts of the plant. Sugarcane is exceptionally tolerant. May be applied by air. The use of a surfactant is recommended on certain crops.

PRECAUTIONS: Do not use with a surfactant on turf and ornamentals.

ADDITIONAL INFORMATION: Signs of herbicidal action are yellowing of leaves, stunting of the plant, and finally death of the growing points. Most rapid activity occurs at high temperatures. Low toxicity to mammals, birds, and fish. Rumex species are especially susceptible to this compound. May be tank mixed with other herbicides. Primarily used to control established perennial grass species. Herbicidal activity may not be visible for 2-3 weeks. Some preemergence activity, but is short lived.

NAMES

PROSULFOCARB, BOXER, DEFI, ARCADE

S-benzyl dipropylthiocarbamate

TYPE: Prosulfocarb is a carbamate compound used as a selective pre and early postemergence herbicide..

ORIGIN: 1985 Stauffer Chemical Co. ICI Ag Products is now the basic producer.

TOXICITY: LD_{50} - 1820 mg/kg. May cause eye and skin irritation.

FORMULATION: 8 EC.

USES: Being sold in Europe for use on cereals. Experimentally being tested on potatoes.

IMPORTANT WEEDS CONTROLLED: Galium spp, chickweed, Alopecurus spp, Poa spp, lambsquarters, mustards.

RATES: Applied at 3-4 kg ai/ha.

APPLICATION: Applied as a preemergence or early postemergence application. Apply postemergence when the weeds are in the early cotyledon to the early branching stage.

PRECAUTIONS: Not for sale or use in the U.S. Moderately toxic to fish. Postemergence applications may cause injury to winter barley. Cereals should be planted to a depth of 12-25 mm for crop safety.

ADDITIONAL INFORMATION: Weeds often emerge but quickly die. There appears to be no varietal differences to the selectiveness of this compound. Short soil persistence. Wild oats are not controlled. Water solubility is 13 ppm.

NAMES

THIOBENCARB, BOLERO, SATURN, SATURNO, SIACARB, TAMARIZ

S[(4-chlorophenyl)methyl] diethylcarbamothioate

TYPE: Thiobencarb is a carbamate-compound used as a postemergence, selective herbicide.

ORIGIN: 1965. Kumiai Chemical Industry Co. Ltd. of Japan. Licensed to be marketed in the U.S. by Valent Chemical Co.

TOXICITY: LD_{50} - 920 mg/kg.

FORMULATIONS: 10% granules, 8EC.

RATES: Applied at 3-6 Ib actual/A.

USES: Rice, celery, endive and lettuce.

IMPORTANT WEEDS CONTROLLED: Barnyardgrass, sprangletop, dayflower, ducksalad, redstem, teaweed, spikerush, water hyssop, junglerice, sedge, signalgrass, tago weed, and other annual weeds.

APPLICATION:

1. Transplant rice—Apply 3 days before until 10 days after transplanting. Also may be incorporated in the soil 1-3 days before transplanting. Hold water at more than 3 centimeters during application. Controls barnyardgrass postemergence up to the 2-leaf stage.

2. Drilled rice—Applied preemergence, or postemergence (up to 15 days after seeding). May be mixed with propanil. May be used on water-seeded rice.

3. Preplant—Used on some row crops as a preplant incorporated herbicide.

PRECAUTIONS: Do not apply to second crop rice. Do not apply within 14 days before or after an organophosphate or carbamate insecticide. Do not apply with liquid nitrogen, zinc or surfactants. Do not apply to exposed rice seed. Toxic to fish and crayfish. Avoid drift.

ADDITIONAL INFORMATION: Remains active in the soil for 30-40 days. Sunlight does not break it down. In water, solubility is 30 ppm. May be mixed with other pesticides. Works well in high organic matter soils as a preplant herbicide on some crops.

HETEROCYCLIC NITROGEN DERIVATIVES

Triazines, Pyridines, Pyridazones, Picolinic Acid, Sulfonylurea Compounds, Imidazole Compounds, etc.

NAMES

AMINOTRIAZOLE, ATA, AMITROLE, AMITROL-T, AMIZOL, AZOLAN, CYTROL, WEEDAZOLE, RADOXONE, AMEROL, AZOLE, MIZOL, SUPERZOL

$$H-N-N$$
$$\underset{N}{\overset{C}{\diagup}}\underset{}{\overset{C}{\diagdown}}-NH_2$$

3-amino-1,2,4-triazole

TYPE: ATA as a triazole compound used as a postemergence, non-selective, translocated herbicide.

ORIGIN: 1953. Amchem Products Co. Being sold today by Rhone Poulenc, Bayer and others.

TOXICITY: LD_{50} - 1100 mg/kg.

FORMULATIONS: 2 SC, 90% SP. Formulated with other herbicides.

USES: Non-crop areas and hardwood nursery stock. Used outside the U.S. as a noncrop herbicide in orchards and vineyards and on fallow land prior to planting.

IMPORTANT WEEDS CONTROLLED: Quackgrass, horsetail, thistles, bermudagrass, nutgrass, leafy spurge, whitelop, dock, cattails, poison oak and ivy, foxtails, watergrass, bluegrass, pigweed, and many others.

RATES: Applied at 1-20 Ib actual/A.

APPLICATION:

Thoroughly cover the foliage without loss by runoff. Cultivation about two weeks after applicahon gives the best results on perennial weeds. Weeds should be growing vigorously at the time of application. Poor results have occurred on drought-stressed or overmatured weeds. If possible, treat before the weeds are 5-6 inches high. Heavy rains following the application will reduce the effectiveness.

PRECAUTIONS: Morningglory or bindweed are considered resistant, as are tules and Russian knapweed. Usage in the U.S. is very limited.

ADDITIONAL INFORMATION: Requires 2-3 weeks for complete results. Hinders or inhibits formulation of chlorophyll so the plant turns white, red, or brown. Accumulates in the meristematic regions. Quickly inactivated in heavy soils, so soil application for root absorption is inadequate. Older plants absorb this material slowly. Persistence in the soil may last for 2 weeks. The activity of the herbicide is enhanced by the addition of ammonium thiocyanate.

NAMES

TRICLOPYR, GARLON , TIMBREL, TOGOR, BASELINE, ZYTRON, REMEDY, RELY, EXETOR, REDEEM, GARIL, GRAZON ET, PATHFINDER, RELEASE, FENEROW, TURFLON ESTER, GRANDSTAND

3,5,6-trichloro-2-pyridinyloxyacetic acid

TYPE: Triclopyr is a picolinic acid compound used as a selective, postemergence herbicide.

ORIGIN: 1973. The Dow Chemical Company (now DowElanco).

TOXICITY: LD_{50} - 713 mg/kg. Irritating to the eyes and skin.

FORMULATIONS: 3 and 4 Ib/gal as the amine salts and ester formulation.

IMPORTANT WEEDS CONTROLLED: Pines, ash, sassafras, hemlock, poison oak, maples, oaks, locust, manzanita, blackberries, cherry, willows, rose, mesquite, and many other perennial weed and brush species.

USES: Grasses on pastures and rangeland and turf. Being used for vegetation management on right-of-ways, industrial and forestry sites. Experimentally being used on small grains and rice and as an aquatic herbicide.

RATES: Applied at .25-4 kg a.i./ha.

APPLICATION: Apply when brush species are actively growing. May also be injected into the tree or applied to freshly-cut stumps to prevent resprouting. To turf apply as a postemergence treatment. Used in conifer release programs.

PRECAUTIONS Avoid drift. Do not plant conifer seedlings for 6 months after application. Do not use in irrigation ditches. Toxic to fish. Avoid drift.

ADDITIONAL INFORMATION: Absorbed by both the leaves and roots and readily translocated throughout the plant. Best results are obtained when the soil moisture is adequate for normal plant growth. More effective than 2,4,5-T on many species. Degrades rapidly in the soil. Low fish toxicity. Most effective on root-suckering species of brush. Established grasses are not injured by rates needed for weed and brush control. Water solubility is 430 ppm. May be mixed with other brush killers. May be aerially applied. Used to suppress bermudagrass and kikyugrass encroachment in cool season turf species.

RELATED MIXTURES:

1. CROSSBOW—A combination of triclopyr and 2,4-D marketed by DowElanco to use on non-crop areas, industrial sites, rangeland and pastures for weed and brush control.

2. TURFLON D and TURFLON II—A combination of triclopyr and 2,4-D developed by DowElanco for use on ornamental turf.

3. ACCESS—A combination of picloram and triclopyr marketed by DowElanco for the control of woody plants in forests, industrial and non crop areas.

4. CONFRONT—A combination of 2.25 lb. triclopyr and .75 lb clopyralid marketed by DowElanco for usage on ornamental turf.

NAMES

FLUROXYPYR, STARANE, ARIANE, ADVANCE, SICKLE, SPITFIRE, TRISTAR, FOLLOW

4-amino-3,5-dichloro-6-fluoro-2-pyridyloxyacetic acid

TYPE: Fluroxypyr is a pyridyl compound used as a selective translocated postemergence broadleaf herbicide.

ORIGIN: 1982. Dow Chemical Co. (now DowElanco).

TOXICITY: LD_{50} - 2405 mg/kg. May cause eye irritation.

FORMULATIONS: 1.67 Ib EC. Formulated with other herbicides.

IMPORTANT WEEDS CONTROLLED: Most broadleaf weeds including wild buckwheat, bedstraw, cleavers, chickweed, bindweed, dock, morningglory, velvetleaf, jimsonweed, oxalis, and many others.

USES: Used outside the U.S. on cereals, corn, orchards, turf, vineyards, pastures, forestry, plantation crops, and other crops.

RATES: Applied at 125-500 g ai/ha.

APPLICATION: Apply as a postemergence spray when the weeds are growing actively. Repeat as necessary. In forestry, used as a foliar spray when conifers are dormant, as a eut surface treatment or as a basal bark treatment. Apply to grain from the 2-leaf stage up to flag stage emergence.

PRECAUTIONS: Not for sale or use in the U.S. Do not apply if rain is expected within 2 hours. Avoid drift. In orchards and vineyards, do not let the spray come in contact with the trees or vines. Toxic to fish.

ADDITIONAL INFORMATION: Readily translocatable. Controls brush species in forestry programs. Fast acting in warm weather. Grasses are very tolerant to this material. May be tank mixed with other herbicides. Low volalility so drift problems are greatly reduced.

NAMES

PICLORAM, AMDON, BORLIN, GRAZON, TORDON

4-amino-3,5,6-trichloro-2-pyridinecarboxylic acid

TYPE: Picloram is a picolinic acid compound used as a selective translocated postemergence herbicide.

ORIGIN: 1963. The Dow Chemical Company. Now marketed by DowElanco.

TOXICITY: LD_{50} - 4012 mg/kg.

FORMULATIONS: 2 and 10% pellets, 2EC, 240 SL. Also formulated in combination with other herbicides.

USES: Non-crop areas, cereal crops, rangelands, forestry and pastures.

IMPORTANT WEEDS CONTROLLED: Field bindweed, Canada thistle, leafy spurge, Russian knapweed, aspen, oak, pine, poison oak, mesquite, and many other broadleaf herbacious and woody plants.

RATES: Applied at .25-8 Ib actual/A.

APPLICATION: May be applied through ground or aerial-application equipment. Special care is required to confine the spray to the target area. Foliage sprays may be applied whenever plants are actively growing. Basal bark treatments can be applied at any time. Apply granules over the roots of plants to be controlled during the normal growing season and when rainfall can be expected soon after treatment. Also used to kill trees by the injection technique.

PRECAUTIONS: Avoid drift. Do not apply under the drip line of desired trees. Do not apply to irrigation ditches. Avoid movement of treated soil. Herbicidally active residues may remain in the soil for a considerable length of time. Very minute quantities will injure many broadleaf crop plants. Do not apply where surface water from the treated area can run off onto cropland. Do not apply to frozen or saturated ground. Do not move livestock from a treated area to any area where some injury cannot be tolerated without first letting them graze on an untreated area for 7 days, as it takes this long for the chemical to leave their urine.

ADDITIONAL INFORMATION: The herbicidal action is a result of absorption through leaves from foliar sprays, uptake through the roots from soil, and residual activity in the soil. The rate required varies according to weed or brush species and geographical location. Higher rates of the granular products are generally required for woody plant control. Established stands of most grasses are tolerant. Seedling grasses are more susceptible. Non-corrosive, non-volatile, and non-flammable.

RELATED MIXTURES:
1. Pathway—A combination of picloram and 2,4-D developed by DowElanco to control unwanted trees.

2. Tordon 101—A combination of picloram and 2,4-D developed by DowElanco for control of woody species.

3. Tordon 1 Plus 2 Mixture—A combination of picloram and triclopyr developed by DowElanco for brush control.

4. Grazon P+D—A combination of picloram and 2,4-D developed by DowElanco to control weeds in pastures and rangelands.

NAMES

CLOPYRALID, LONTREL, LONTRYX, RECLAIM, SHIELD, STINGER, MATRIGON, TRANSLINE

3,6-dichloro-2-pyridinecarboxylic acid

TYPE: Clopyralid is a picolinic-acid derivative that is being used as a postemergence, selective herbicide.

ORIGIN: 1975. Dow Chemical Company. Now marketed by DowElanco.

TOXICITY: LD_{50} - 4300 mg/kg. A severe eye irritant.

FORMULATIONS: 3 EC.

USES: Wheat, oats, barley, sugar beets, grass crops, corn, forestry, Christmas trees, turf, pastures, fallow cropland, non-cropland areas, and on rangeland. Used outside the U.S. on these crops as well as flax, onions, strawberries and cole crops.

IMPORTANT WEEDS CONTROLLED: Canadian thistle, Russian thistle, sunflower, wild buckwheat, smartweed, cocklebur, knapweed, mayweed, ragweed, sowthistle and others. Many broadleaf weeds, especially Composites, Polygonacea, legumes and woody plants such as mesquite and acacias.

RATES: Applied at 50-200 g a.i./ha.

APPLICATION: Applied as a postemergence application when weeds are small. Apply to sugar beets in the 2-8 leaf stage. Apply to corn from emergence to 24 inches tall. On small grain apply from the 3 leaf stage to the early boot stage.

PRECAUTIONS: Avoid drift. Do not use in irrigation ditches. Do not spray pastures containing desirable forbs. Soil temperature below 70°F results in poor control. Do not apply to bentgrass turf. Do not apply by air.

ADDITIONAL INFORMATION: Readily absorbed by leaves and roots and translocated throughout the plant. May be tank mixed with other herbicides. Grasses are tolerant. Only weeds that have emerged at time of application will be controlled.

RELATED MIXTURES:

1. CURTAIL—A combination of clopyralid and 2,4-D sold by DowElanco for use on small grains to control broadleaf weeds. Also used on grasses grown for seed.

2. CURTAIL-M—A combination of clopyralid and MCPA developed by DowElanco for use on small grains. Also used on grasses grown for seed.

NAMES

DIFENZOQUAT, AVENGE, FINAVEN, SUPERAVEN

1,2-dimethyl-3,5-diphenyl-1H-pyrazolium

TYPE: Difenzoquat is a pyrazolium compound used as a selective, postemergence

ORIGIN: 1972. American Cyanamid Company.

TOXICITY: LD_{50} - 270 mg/kg. May cause eye and skin irritation.

FORMULATION: 2EC.

RATES: Applied at .5-1.4 kg a.i./ha.

USES: Barley and wheat.

IMPORTANT WEEDS CONTROLLED: Wild oats.

APPLICATION: Apply postemergence when the wild oats are in the 3-5-leaf stage.

PRECAUTIONS: Avoid drift and overlapping. Wheat varieties differ in their tolerance to this material . Do not graze treated area. Do not apply when plants are wet with heavy dew or rain or when rain is predicted within a minimum of 6 hours. Corrosive to aluminum.

ADDITIONAL INFORMATION: Compatible with other herbicides that control broadleaf weeds. May be applied by air. Controls by contact only. No preemergence activity.

NAMES

PYRIDATE, KANIDE, FENPYRATE, LENTAGRAN, TOUGH

0(6-chloro-3-phenyl-4-pyridazinyl)S-octyl-carbonothioate

TYPE: Pyridate is pyridazine compound used as a selective, contact, postemergence herbicide.

ORIGIN: 1976. Chemie Linz AG of Austria. Being developed in the U.S. by Agrolinz and Helena Chemical Co.

TOXICITY: LD_{50} - 4600 mg/kg. May cause some eye and skin irritation.

FORMULATIoN: 45% WP. 3.75 EC.

USES: Experimentally being tested in the U.S . on alfalfa, peanuts, corn, cotton and vegetables. Used in Europe on rice, corn, brassicas, cereals, and other crops.

IMPORTANT WEEDS CONTROLLED: Galium spp., Galeopsis spp., Lamium spp., Veronica spp., barnyardgrass, crabgrass, foxtails, lambsquarters, nightshade, panicums, pigweed, and others.

RATES: Applied at 1-1.5 kg a.i./ha, or at .9-1.8 Ibs ai/acre.

APPLICATION: Applied postemergence when the weeds are no larger than the 4-leaf stage.

PRECAUTIONS: Used on an experimental basis only in the U.S. Perennial weeds are not controlled. Application during hot weather may cause a slight burning of the cereal leaves. Do not use with wetting agents. Allow at least 1 week between treatment and the use of 2,4-D or fertilizer sprays.

ADDITIONAL INFORMATION: Water solubility is 90 ppm. Does not carry over in the soil to harm following crops. Rainfall after treatment does not diminish the degree of control. Slow acting. Rapidly decomposes in the soil so there is no preemergence activity, nor any possibility of carryover. Death to the weeds occur within 5-7 days. May be tank mixed with other herbicides to increase the spectrum of control.

NAMES

PYRAZON, CHLORIDAZON, BETAZON, CLORIPUR, CURBETAN, PYRAMIN, LENAPAC, HYZON, TROJAN, BONUS, SUZON, SILVER, BONUS, PIRAMIN

5-amino-4-chloro-2-phenyl-3(2H)-pyridazinone

TYPE: Pyrazon is a pyridazone compound used as a selective, pre and postemergence herbicide.

ORIGIN: 1962. BASF AG. of Germany.

TOXICITY: LD_{50} - 2140 mg/kg. May be irritating to the skin.

FORMULATIONS: 65%DF, 4.2 lb/gal F. Formulated with other herbicides.

USES: Sugar beets and red beets.

IMPORTANT WEEDS CONTROLLED: Lambsquarters, mustards, ragweed, purslane, smartweed, pigweed, henbit, shepherd's purse, nightshade, groundsel, dock, spurge, chickweed, wild radish, and others.

RATES: Applied at 2-4 Ib ai/A.

APPLICATION:

Apply preplant incorporated or preemergence. In areas of reliable rainfall or sprinkler irrigation, it should not be incorporated. Incorporate by power-driven rotary tillers or similar devices in arid regions to a depth of two inches under furrow irrigation only. Irrigate within 3 days of treatment. Apply after beets have 2-5 leaves, but before weeds have formed 2-4 true leaves. Postemergence control of pigweed has been variable. Timing is critical so do not use on weeds larger than the 2-4-leaf stage.

PRECAUTIONS: Do not use on sands, loamy sands, muck, or peat soils. Do not sprinkler irrigate if used as a soil-incorporation treatment. Do not incorporate overall with a disc. Cabbage, carrots, cucumbers, lima beans, and tomatoes are especially susceptible to this material. Weeds lose much of their susceptibility by the time they have 4 true leaves. Do not apply when temperature and humidity are high or when beet foliage is wet. Perennial weeds are not controlled. Preemergence applications are not effective under furrow irrigation unless incorporated. Thorough agitation is necessary.

ADDITIONAL INFORMATION: Fairly resistant to leaching. Water solubility is 400 ppm. Emerged crop showed more tolerance than when put on preemergence. Sugar beets are susceptible during the cotyledon stage; therefore, do not apply during this period. Often combined with other pesticides and herbicides. Moisture is required to activate this material. Weed control can be expected for 4-8 weeks. Somewhat ineffective on grasses. A light incorporation (1-2 inches) gives better results than a deep one (4-5 inches). Weeds absorb this material primarily through their root systems. Non-corrosive. Use on soils where the organic matter content is higher than 5% will result in erratic weed control under dry weather conditions.

NAMES

PYRAZOSULFURON-ETHYL, NC-311, SIRIUS, KALCRON,
INAZUMA, ALT, SURVEYOR, BOLSAR, RISER, BERUF, AGREEN

Ethyl 5-[3-(4,6-dimethoxypyrimidim-2-yl)-carbonoylsulfamoyl)-1-
methylpyrazole-4-carboxylate

TYPE: Pyrazosulfuron-ethyl is a sulfonylurea compound used as a selective pre and postemergence herbicide.

ORIGIN: 1982. Nissan Chemical Industries of Japan.

TOXICITY: LD_{50} - 5000 mg/kg.

FORMULATIONS: .07% granules, 25% F, 5% WP. Formulated with other herbicides.

USES: Outside the U.S. on rice and turf.

IMPORTANT WEEDS CONTROLLED: Barnyardgrass, Cyperus spp, Monochoria, Scripus spp. and others

RATES: Applied at 10-40 g ai/ha.

APPLICATION: Applied from 1-20 days after transplanting. On direct seeded rice apply when in the 1-3 leaf stage.

PRECAUTIONS: Not for sale or use in the U.S.

ADDITIONAL INFORMATION: Absorbed mainly through the root system. Controls annual and perennial broadleaf weeds and sedges as well as barnyardgrass. Highly selective to any rice variety. Being used on both transplant and direct seeded rice.

NAMES

THIAZAFLURON, ERBOTAN

1,3-dimethyl-1-(5-trifluoromethyl-1,3,4-thiadiazol-2-yl)urea

TYPE: Thiazafluron is a urea compound used as an early-post- and preemergence herbicide.

ORIGIN: 1972. CIBA-Geigy Ltd.

TOXICITY: LD_{50} - 278 mg/kg. May cause eye irritation.

FORMULATIONS: 50% and 80% WP, 10% granules.

IMPORTANT WEEDS CONTROLLED: Annual and perennial broadleaves and grasses.

USES: Outside the U.S. for industrial weed control.

RATES: Applied at 2-12 kg actual/ha.

APPLICATION: Applied while the weeds are dormant for the best results. Moisture must carry it into the rootzone.

PRECAUTIONS: Not for sale in Ihe U.S. Do not apply near desired plants.

ADDITIONAL INFORMATION: 2100 ppm. water solubility. Mainly active through the plant root system. Poor foliar activity. May be mixed wilh other herbicides. Long lasting, often giving 2-3 years of control.

PIPEROPHOS, RILOF

O,O-dipropyl S-[2-(2'-methyl-piperidinocarbonyl] methyl phosphorodithioate

TYPE: Piperophos is a organic phosphate compound used as both a pre and postemergence herbicide.

ORIGIN: 1969. ClBA-Geigy Ltd.

TOXICITY: LD_{50} - 324 mg/kg. May cause eye and skin irritation.

FORMULATIONS: 50%EC. Also formulated with dimethametryn under the trade name Avirosan and with 2,4-D which is trademarked RILOF-H. These are both used on transplant rice.

IMPORTANT WEEDS CONTROLLED: Annual grasses and sedges.

USES: Outside the U.S. on seeded or flooded rice.

RATES: Used at 2-3 kg ai/ha.

APPLICATIONS: Applied as a preemergence treatment.

PRECAUTIONS: Not for sale in the U.S.

ADDITIONAL INFORMATION: Taken up by young plants through roots, coleoptiles, and leaves. 25 ppm water solubility. Mainly used with other herbicide.

NAMES

NORFLURAZON, EVITAL, SOLICAM, TELOK, ZORIAL, PREDICT

[Chemical structure diagram: a trifluoromethyl-substituted phenyl ring connected to a pyridazinone ring bearing a chloro substituent, a ketone (O), and a methylamino (N–CH₃, H) group]

4-chloro-5-(methylamino)-2(alpha, alpha, alpha-trifluoro-m-tolyl)-3-(2H)-pyridazione

TYPE: Norflurazon is a pyridazinone compound used as a selective preemergence herbicide.

ORIGIN: 1971. Sandoz Ltd. of Switzerland.

TOXICITY: LD_{50} - 8000 mg/kg.

FORMULATIONS: 80 WP, 5% granules.

USES: Cotton, almonds, apples, citrus, asparagus, ornamentals, blueberries, caneberries, grapes, soybeans, hops, pecans, cranberries, apricots, cherries, filberts, nectarines, peaches, plums, prunes, walnuts, and for non-crop weed control.

IMPORTANT WEEDS CONTROLLED: Control of annual grasses, fall panicum, spikerush, saltgrass, morningglory, pigweed, prickly sida, ragweed, cheeseweed, purslane, fiddleneck, shepherd's purse, and others.

RATES: Applied at 1-8 Ib active/A.

APPLICATION: Applied to the soil surface prior to weed emergence, either band or broadcast. Also can be applied to cotton preplant incorporated followed by a preemergence spray later in the season. Apply to perennial crops while they are dormant. Sprinkler irrigation is recommended if no rain occurs within 2 weeks of application. May be soil incorporated. Apply to cranberries in the early spring or in the fall after harvest.

PRECAUTIONS: Do not graze treated areas. Avoid drift. Major crops that show no tolerance to this herbicide are alfalfa, beans, cole crops, corn, small grains, rape, peas, rice, tomatoes, sugar beets, and others. If stand is lost, do not replant to another crop other than cotton, soybeans, or peanuts. Apply to tree crops that have been established at least one year.

ADDITIONAL INFORMATION: Water solubility is 23 ppm. Relatively low toxicity to fish and birds. A light incorporation does not reduce the activity. Treated acreage can be rotated to cotton, soybeans, or peanuts in 12 months, except on heavy soil where a delay of 18 months is required. Agitate when applying. Absorbed through the plant's root system.

NAMES

DIQUAT, **ACTOR, AQUACIDE, DEXTRONE, MIDSTREAM, PREEGLONE, PRIGLONE, REGLONE, REGLOX, REGLEX, TORPEDO**

6,7-dihydroipyridol[1,2-a:2',1'-c]pyrazidinium dibromide

TYPE: Diquat is a contact, non-selective herbicide and plant desiccant, applied postemergence.

ORIGIN: 1958. ICI Ltd. (Europe). Licensed to be sold in the U.S. by Valent Chemical Company.

TOXICITY: LD_{50} - 231 mg/kg. May be irritating to the skin and eyes.

FORMULATIONS: 2 AS. Formulated with other herbicides.

USES: A preharvest desiccant on clover, soybeans, potatoes, vetch, and alfalfa seed crops. Also used as a general-contact weed killer and on aquatic weeds. Used in Europe as a precrop emergence herbicide and as a cotton and potato desiccant. On sugarcane, used for weed control and for tassel inhibition.

IMPORTANT WEEDS CONTROLLED: Practically all annual plants that it comes in contact with, as well as aquatic weeds. Annual grasses are less susceptible than broadleaf plants.

RATES: Applied at 1-2 Ib actual/A.

APPLICATION: Should be mixed with a spreader. Most effective when applied in the evening or at night. Thoroughly cover the plant foliage. Repeat as necessary. For submerged aquatics, apply to the waler surface or inject below the surface and distribute as evenly as possible. Repeat when new infestation occurs. Do not apply to muddy water. Often applied in combination with copper compounds. Treated water can be used for irrigation and other purposes. For sugarcane, apply during flower initiation for tassel inhibilton.

PRECAUTIONS: Do not apply to crops that are wet from rain. Do not apply near desired plants. Do not use treated water for spraying or overhead irrigation for 14 days following treatment. To avoid drift, do not apply during periods of thermal inversion or windy conditions. Corrosive to most metals.

ADDITIONAL INFORMATION: Effective at very low dosages. Rapid acting with visible effects in just a few days. Inactivated immediately upon contact with the soil. Not effeclive on perennial weeds, since they just grow back after being burned off. Translocated to a limited extent. Non-toxic to fish at the recommended rates. More effective on broadleaves than grasses. Non-volatile. Germination is not reduced by usage of this material on seed crops. Bark of trees completely void of chlorophyll are unharmed by spray so it is useful around woody ornamentals. Rapidly absorbed by the aerial parts of plants.

NAMES

PARAQUAT, CEKUQUAT, CRISQUAT, CYCLONE, GRAMOXONE-EXTRA, PARACOL, SCYTHE, SPEEDWAY, STARFIRE, SPEEDER, PROTEX, RADEX, SIPQUAT, VIOLAN, PREEGLONE, OSAQUAT, ARIAL

1:1-dimethyl-4,4'-bipyridinium dichloride

TYPE: Paraquat is a non-selective herbicide with fast-acting contact action, applied postemergence.

ORIGIN: 1958. In Europe by ICI Ltd.

105

TOXICITY: LD$_{50}$ - 150 mg/kg. Do not breathe the spray mist. Will kill if swallowed.

FORMULATIONS: 2 AS, 2.5 AS, 1.5 AS. Formulated with other herbicides.

USES: Alfalfa, almonds, apples, apricots, asparagus, avocadoes, bananas, beans, cabbage, carrots, cauliflower, cherries, Chinese cabbage, citrus, coffee, collards, com, cotton, cucurbits, figs, filberts, grapes, guar, hops, kiwi, lettuce, melons, nectarines, peaches, peanuts, pears, peas, peppers, pistachios, plums, potatoes, prunes, sugarcane, sorghum, turnips, walnuts, non-croplands, grass seed, cotton defoliant, potato vine desiccant, ornamental trees, macadamia nuts, olives, papayas, peppers, soybeans, sugarbeets, and tomatoes. Used in other countries on rubber trees, palms, and other tropical crops.

IMPORTANT WEEDS CONTROLLED: Most annual weeds and grasses.

RATES: Applied at 140-2210 g a.i./ha.

APPLICATION:

1. Crops and General Weed Control—Apply as a shielded, directed spray, protecting crops from the drift. Spray, hitting woody stems and tree trunks does not cause injury. Repeat application as necessary. Young, succulent weeds are more easily killed. On open cropland, use as a preplant or preemergence spray. Widely used in no-till or minimum-tillage programs.

2. Desiccation and Defoliation—Cover foliage thoroughly. If growth is extremely heavy, two applications are advisable. Do not apply within three days of harvest.

PRECAUTIONS: Regrowth may occur in perennials with large, underground root systems. Do not use with anionic surfactants. Use within a few hours of mixing. Avoid drift.

ADDITIONAL INFORMATION: Immediately inactivated when in contact with the soil, thereby building up no residues. Rapid acting. Non-volatile. Can be translocated, but most activity is by local absorption. Use with a non-ionic surfactant. May be combined with soil sterilants for fast knockdown effects. Corn and sorghum have not been injured by a directed, non-shielded spray in the row and on the base of the plants after they had reached 18 inches in height. Used in range improvement programs and on aquatic weeds. Bark of trees, once free of chlorophyll, will not be injured. Compatible with many other residual herbicides. Rainfall after application has little bearing on the effectiveness. Weeds should be 1-6 inches high for the best results. Low fish toxicity.

RELATED MIXTURES:

Surefire—A combination of paraquat and diuron developed by Platte Chemical Co. for use as a fallow-land herbicide or for noncrop weed control.

NAMES

SIMAZINE, **BATAZINA, BITEMOL, CEKUZINA-S, SIM-TROL, GESATOP, HERLAZIN, PRIMATOL S, PRINCEP, PRINTOP, SIMANEX, ZEAPUR, SIM-TROL, WEEDEX, LUSERB, SIMFLOW**

$$CH_3 — CH_2 — N — C \underset{\underset{N}{\|}}{\overset{\overset{Cl}{\underset{\|}{C}}}{\overset{N}{\underset{N}{\overset{\|}{\ }}}}} C — N — CH_2 — CH_3$$

6-chloro-N', N'-diethyl-1,3,5-triazine-2,4-diamine

TYPE: Simazine is a selective triazine compound used as a preemergence herbicide.

ORIGIN: 1956.ClBA-Geigy Corp. Produced by a number of other manufacturers today.

TOXICITY: LD_{50} - 5000 mg/kg. May cause slight eye and skin irritation.

FORMULATIONS: 80% WP; 90 DF, granules, 4F.

USES: Apples, avocados, almonds, pecans, turf, blueberries, strawberries, filberts, dewberries, plums, almonds, currants, peaches, caneberries, citrus, corn, cranberries, grapes, macadamia nuts, olives, pears, pineapples, cherries, walnuts, and ornamentals. Also used as an aquatic herbicide outside the U.S.

IMPORTANT WEEDS CONTROLLED:

Barnyardgrass, mustard, chickweed, crabgrass, algae, foxtails, jimsonweed, lambsquarters, purslane, ragweed, pigweed, Russian thistle, wild oats, velvetleaf, and many others.

RATES: Applied al 1-40 Ib actual/A.

APPLICATION: Applied as a preemergence herbicide. Rainfall is required to move it into the soil.

PRECAUTIONS: Mission variety of almonds may be injured. Grapes must be 3 years old before application is made. Sensitive crops include tomatoes, tobacco, oats, spinach, cucurbits, onions, clovers, carrots, rice, beets, soybeans, crucifers, and lettuce. Long residual effects. Do not plant another crop the same season or in the fall as injury may result. Do not apply to trees growing in sandy soils. Do not apply to nut crops when nuts are on the ground. Make only 1 application to trees per year. Use on turf only in areas designated on the label. No longer used in the U.S. for non crop weed control.

ADDITIONAL INFORMATION: Does not prevent germination, but destroys seedling after entering the roots. Little, if any, foliage contact action. Corn metabolizes it, making it resistant. Tightly held by the soil. Microorganisms break it down in about a year. Persists longer in dry, cold, or low-fertility soils. Plowing will reduce the possibility of injury to following crops. Broadleaf plants are the most susceptible. Non-flammable. Rainfall is required to activate the chemical. Under dry conditions, a shallow soil incorporation gives better weed control. Water solubility is about 5 ppm. Low toxicity to fish and wildlife. Mixed with other herbicides.

RELATED MIXTURES:

SIMAZAT—a simazine/atrazine combination developed by Drexel Chemical Co. for use on corn, sugarcane and Christmas trees.

NAMES

ATRAZINE, AATREX, **AATREX-NINE-O, CHEAT-STOP, ATAZINAX, ATRANEX, CANDEX, CEKUZINA-T, GESAPRIM, INAKOR, PRIMAZE, ATRADEX, RADAZIN, ZEAPOS, ZEAZIN, FOGARD, WEEDEX, MALERMAIS**

6-chloro-N-ethyl-N'-(1-methylethyl)-1,3,5-triazine-2,4-diamine

TYPE: Atrazine is a selective triazine herbicide used preemergence and early postemergence.

ORIGIN: 1959. CIBA-Geigy Corp. Produced today by a number of manufacturers.

108

TOXICITY: LD$_{50}$ - 1750 mg/kg. May cause slight eye and skin irritation.

FORMULATIONS: 80% WP, 4F, 90DF, granules. Formulated with other herbicides.

USES: Corn, guava, macadamia orchards, non-crop areas, pastures, sugarcane, sorghum, turf grass sod, forestry and Christmas tree plantations. Used on these and many other crops outside the U.S.

IMPORTANT WEEDS CONTROLLED:

Barnyardgrass, mustards, chickweed, cocklebur, crabgrass, downy brome, foxtail, jimsonweed, lambsquarters, nutgrass, quackgrass, purslane, ragweed, velvetleaf, wild oats, and many others.

RATES: Applied at 1-60 lb a.i./acre.

APPLICATION:

Applied pre or postemergence when weeds are less than 1 1/2 inches tall. When applied to larger weeds, the broadleaves are controlled satisfactorily, but results with grasses are variable. Agitate while spraying. Soil incorporation gives good results in irrigated areas.

PRECAUTIONS: Sensitive crops include most vegetables, cereal grains, asparagus, soybeans, peanuts, and potatoes. Residues may remain in the soil for over one year. May move laterally with water. Also leaches downward. Do not apply near desired plants. In the U.S. do not apply to corn or sorghum at rates of over 3 lb a.i./year.

ADDITIONAL INFORMATION: Moisture activates the chemical. Usually considered more toxic than the other triazines. Non-flammable. Will not control johnsongrass or bermudagrass. If oats follow corn, the soil should be tilled in the fall to minimize possible injury. Acts mainly through the roots, but there is some activity through foliage contact. Corn completely metabolizes it, not being injured. Effective on most annual broadleaves for about three months. Resembles simazine, but is faster acting under low-rainfall conditions. Does not prevent germination, but kills weeds after being absorbed by the root system. A shorter residual life than simazine. Agitate while spraying. Proven and used in all climates. Aerial application has proven successful. Solubility in water 33 ppm. Mixed with other herbicides.

NAMES

PROMETON, GESAGRAM, PRAMITOL, PRIMATOL, GESAFRAM

6-methoxy-N',N'-bis(1-methylethyl)-1,3,5-triazine-2,4-diamine

TYPE: Prometon is a non-selective triazine herbicide which is applied both pre and postemergence.

ORIGIN: 1957. CIBA-Geigy Corp.

TOXICITY: LD_{50} - 2100 mg/kg. May cause eye irritation.

FORMULATIONS: 2EC, granules. Formulated with other herbicides.

USES: Total vegetation control and brush control in non-crop areas.

IMPORTANT WEEDS CONTROLLED: Johnsongrass, bermudagrass, foxtall, mustards, ragweed, plantain, quackgrass, horsetail, watergrass, chickweed, cocklebur, mullein, crabgrass, dock, goosegrass, jimsonweed, lambsquarters, nightshade, puncturevine, purslane, pigweed, velvetleaf, wild oats, and many others.

RATES: Applied at 10-60 Ib actual/A.

APPLICATION: Applied either at time of weed emergence or 2-3 months afterwards. May be applied in oil.

PRECAUTIONS: Do not use near desired plants. Agitate while spraying. Do not use on cropland.

ADDITIONAL INFORMATION: Photosynthetic poison. Faster acting than simazine, but does not have as long a residual effect as either simazine or atrazine. Controls for a full season or longer. Requires rainfall to move it into the soil. Acts through both the roots and the foliage. Non-corrosive to metals. Low toxicity to birds and fish. Also used under asphalt.

RELATED MIXTURES:

1. PRAMITOL 5P.—An industrial, non-selective soil herbicide containing 5% prometon, .75% simazine, 40% sodium chlorate, and 50% sodium metaborate, developed by CIBA-Geigy Corp.

NAMES

PROPAZINE, GESAMIL, MILOGARD, PROZINEX, MILO-PRO, PROZINEX

6-chloro-N', N'-bis (1-methylethyl)-1,3,5-triazine-2,4-diamine

TYPE: Propazine is a selective triazine herbicide applied preemergence, giving residual control.

ORIGIN: 1957. CIBA-Geigy Corp.

TOXICITY: LD_{50} - 7000 mg/kg. May cause slight eye and skin irritation.

FORMULATIONS: 4 F, 50 and 80% WP, 90 WDG.

USES: Sorghum, celery, carrots and parsley outside the U.S.

IMPORTANT WEEDS CONTROLLED: Morningglory, carpetweed, lambsquarters, pigweed, ragweed, foxtails, smartweed, velvetleaf, and many others.

RATES: Applied at .5-3 kg a.i./ha.

APPLICATION: Applied at the time of planting or immediately after planting, before weeds and sorghum emerge. Use either broadcast or band treatment, applying to the soil surface. Rains will then move it into the root zone where the germinating weeds will absorb it. May be soil incorporated where low rainfall conditions prevail. Requires agitation during application. Provides a full season's control.

PRECAUTIONS: Do not plant any other crop for at least 18 months, except corn, cotton, or soybeans which may be planted in 12 months. Do not apply to sandy soils. Sugar beets and many vegetables are very sensitive to this material. Poor control is obtained on Panicum, Setaria, and Oxalis species. No longer sold in the U.S.

ADDITIONAL INFORMATION: Soon after weeds emerge, the first symptom noticed is a yellowing of the leaf tips and margins which progresses rapidly until the entire plant is killed. The residue remaining in the soil will control the succeeding germinating weeds. Drift is not harmful to plant foliage. Noncorrosive. Moisture is required to move the chemical into the root zone. Used under all climatic conditions. Compatible with pesticides and fertilizers. No postemergence activity. Low toxicity to fish and wildlife. Taken up through the roots only. Water solubility is 5 ppm.

NAMES

PROMETRYN, CAPAROL, COTTON-PRO, GESAGARD, MERKAZIN, POLISIN, PROMETREX, SELEKTIN

$$
CH_3 - S - C \underset{N}{\overset{N}{\rightleftharpoons}} \begin{array}{c} NH-CH< \begin{array}{c} CH_3 \\ CH_3 \end{array} \\ C \\ C-NH-CH< \begin{array}{c} CH_3 \\ CH_3 \end{array} \end{array}
$$

N', N'-bis (1-methylethyl)-6-(methylthio)-1,3,5-triazine-2,4-diamine

TYPE: Prometryn is a triazine compound, used as a selective, pre and postemergence herbicide.

ORIGIN: 1962. CIBA-Geigy Corp.

TOXICITY: LD_{50} - 1802 mg/kg. May cause eye and skin irritation.

FORMULATIONS: 80% WP. 4L. Formulated with other herbicides.

USES: Cotton and celery. Outside the U.S. on these crops plus peas, potatoes, carrots, garlic, leeks, lentils, celery, and sunflowers.

IMPORTANT WEEDS CONTROLLED: Pigweed, crabgrass, mustards, ragweed, smartweed, panicum, lambsquarters, malva, foxtails, barnyard grass, nightshade, wild oats, morningglory, teaweed, ground cherry, purslane, and most other annual weeds.

RATES: Applied at .5-3.2 lb actual/A.

APPLICATION: Apply either as a band or broadcast treatment, before weed seedlings emerge. Use heavier rates on heavier soils. As a layby or postemergence spray, apply before weeds are 2 inches high and after cotton is at least 12 inches high. Use in this manner as a directed spray. A surfactant may be added for postemergence application. Also used as a preplant treatment in some areas. Apply postemergence to celery when crop has 2-5 true leaves or within 2-6 weeks after transplanting.

PRECAUTIONS: Used as a directed spray on cotton while it is young to avoid any foliar injury. Do not use preemergence on light, sandy soils.

ADDITIONAL INFORMATION: Water solubility is 40 ppm. Soil incorporation is not necessary under most conditions. No harmful effects have been noticed when it is sprayed on the lower cotton bolls. Rainfall or sprinkler irrigation will move this material into the soil. Non-corrosive. Absorbed by the roots, as well as giving a foliar kill. More effective on broadleaves than grasses. Does not have the long residual effects that characterize many triazine materials. Weeds will emerge before dying. Agitate while spraying.

NAMES

AMETRYN, AMETREX, EVIK, GESAPAX, AMEFLO

N-ethyl-N'-(1-methylethyl)-6-(methylthio)-1,3,5-triazine-2,4-diamine

TYPE: Ametryn is a selective triazine compound, used as a pre and postemergence herbicide.

ORIGIN: 1964. CIBA-Geigy Corp.

TOXICITY: LD$_{50}$ - 1000 mg/kg.

FORMULATION: 80% WP. Formulated with other herbicides.

USES: Bananas, corn, pineapple, sugarcane, and non-cropland. Used as a vine desiccant on dry beans and potatoes. Outside the U.S. use on these plus citrus, coffee, tea, cocoa, potatoes and oil palms.

IMPORTANT WEEDS CONTROLLED: Crotalaria, mustard, dallisgrass, cocklebur, lambsquarters, morningglory, velvetleaf, ragweed, panicum, shattercane, nutgrass, wiregrass, goosegrass, crabgrass, sowthislle, purslane, pigweed, foxtail, and many others.

RATES: Applied al 1.5-6 kg ai/ha.

APPLICATION: Applied as a preemergence herbicide with some early postemergence activity. Rainfall is required to take it into the soil. Used in some areas as a vine desiccant on potatoes applied 10-14 days prior to harvest.

PRECAUTIONS: Do not apply near desired plants. The chemical moves both vertically and laterally in the soil due to its high water solubility. Do not apply over the top of corn.

ADDITIONAL INFORMATION: Absorbed through the weeds' root system as they germinate. Therefore, weeds will emerge before dying. Considerable activity through foliage contact. Non-corrosive and non-flammable. Mature weeds of certain species will be controlled by postemergence applications. Agitate while spraying. Solubility in water 190 ppm. Compatible with pesticides and liquid fertilizer.

NAMES

DESMETRYN, SEMERON

$$S\text{—}CH_3$$

(chemical structure of 2-isopropylamino-4-methylamino-6-methylthio-s-triazine)

2-isopropylamino-4-methylamino-6-methylthio-s-triazine

TYPE: Desmetryn is a selective triazine compound used as either a pre or postemergence herbicide.

ORIGIN: 1962. ClBA-Geigy Corp.

TOXICITY: LD_{50} - 1390 mg/kg.

FORMULATION: 25% WP.

USES: Outside the U.S. on kale, rape, Brussels sprouts, and cabbage.

IMPORTANT WEEDS CONTROLLED: Fat hen, annual bluegrass, nettle, lambsquarters, pigweed, chickweed, spreading orache, spurrey, nightshade, and others.

RATES: Applied at 500 g a.i./ha.

APPLICATION: The crop should be 5 inches high before spraying and have at least 3 true leaves. Susceptible weeds should be sprayed when they are 2-4 inches high. However, fat hen and spreading orache will be controlled up to 14 inches high. Transplants should not be sprayed for at least two weeks after transplanting.

PRECAUTIONS: Not for sale in the U.S. Rain within 24 hours will reduce the effectiveness. Do not spray when dew is on the leaves. Do not spray poorly-growing crops. Grasses, shepherd's purse, spurge, and pennycress are not controlled. Do not use on broccoli or cauliflower.

ADDITIONAL INFORMATION: Cold weather may reduce the weed control obtained. Slight scorching and loss of color may occur to the crop after spraying, but these effects are quickly outgrown. Effects on weeds are not seen for 7-14 days. Kills by absorption through the roots, as well as foliage contact. Primarily used as a postemergence spray.

Up to 5 weeks weed control can be expected. Water solubility is 600 ppm. Most effective on Chenopodium species.

NAMES

TERBUTRYN, CLAROSAN, IGRAN, **PREBANE, TERBUTREX, PLANTONIT**

$$S-CH_3$$

CH_3 — CH_2 — N — C — C — N — C

N-(1,1-dimethylethyl)-N'-ethyl-6-(methylthio)-1,3,5-triazine-2,4-diamine

TYPE: Terbutryn is a triazine compound being used as a selective, preemergence herbicide.

ORIGIN: 1965. CIBA-Geigy Corp. No longer sold in the U.S.

TOXICITY: LD_{50} - 2000 mg/kg. May cause slight eye and skin irritation.

FORMULATIONS: 50 and 80% WP.

USES: Outside the U.S. it is used as an aquatic herbicide, and on cereals, sugarcane, pineapple, bananas, coffee, tea, cocoa, palms, corn and non crop areas.

IMPORTANT WEEDS CONTROLLED: Fiddleneck, chickweed, mustards, penneycress, knotweed, false flax, henbit, shepherd's purse, prickly lettuce, corn spurry, filaree, mayweed, pigweed, kochia, ragweed, groundcherry, morningglory, lambsquarters, velvetleaf, and others.

RATES: Applied at .8-4 actual/A.

APPLICATION: Apply to the soil surface after the crop has been planted, either before or after emergence of small grains. For post emergence application, apply after the wheat has reached the 3-leaf-stage when it has no more than 1-2 tillers. Weeds should be less than 4 inches in height. Preemergence application depends upon moisture to move it into the soil. May be applied by air. To sorghum, the soil temperatures should have reached 60°F for 3 consecutive days prior to application. Used for total vegetation control where short-term effects are desired, but not permanent sterilization.

PRECAUTIONS: Not for sale or use in the U.S. Some crops in rotation may be injured. Overhead sprinkling to sorghum may injure the crop. Do not apply postemergence when temperatures exceed 70°F. Do not apply if dew is on the leaves. Do not apply more than once per crop cycle. Toxic to fish. Do not apply to emerged sorghum. Do not apply to Sudan-Sorghum hybrids, sorghum breeding stock, or millets.

ADDITIONAL INFORMATION: Absorbed by both the foliage and roots. Being used in combination with propazine or atrazine on sorghum. May be applied with liquid fertilizer. Solubility in water is 25 ppm. Low toxicity to birds.

NAMES

DIPROPETRYN, SANCAP

$$S-CH_2-CH_3$$

```
  CH₃              N   N              CH₃
   |              /     \              |
   CH — NH ——————(       )—— NH — CH
   |              \     /              |
  CH₃               N                CH₃
```

6-(ethylthio)-N', N'-bis (1-methylethyl)-1,3,5 triazine-2,4-diamine

TYPE: Dipropetryn is a triazine compound used as a selective, preemergence herbiclde.

ORIGIN: 1967. ClBA-Geigy Corporation. No longer sold in the U.S.

TOXICITY: LD_{50} - 3900 mg/kg. May cause skin irritation.

FORMULATIONS: 80% WP.

USES: Outside the U.S. it is used on cotton and cucurbits.

IMPORTANT WEEDS CONTROLLED: Pigweed, lambsquarters, groundcherry, barnyardgrass, Russian thistle, crabgrass, sandbur, and others.

RATES: Applied at 1.25-3.5 kg ailha.

APPLICATION: Apply at planting time, or within 2 days after planting, before weeds and cotton emerge. If no rains occur to move the chemical into the root zone, a light incorporation into the soil would be beneficial.

PRECAUTIONS: Rotational crops may be planted 6 months following application. Only banded applications can be made to furrow-planted cotton. Injury to cotton may occur if applied to alkali soils, caliche out-croppings or exposed calcareous soils. Johnsongrass, Coloradograss, bermudagrass, silverleaf nightshade, and groundcherry are not controlled. Do not use on heavy soils. Do not apply to emerged cotton. Toxic to fish. Not for sale or use in the U.S.

ADDITIONAL INFORMATION: Must be absorbed by the weeds to control them . If a planting failure occurs, you can replant into the treated area. Most effective on light, sandy soils. Water solubility is 16 ppm. Herbicidal activity is closely related to prometryn, but less active on a pound-for-pound basis. Not effective as a postemergence application.

NAMES

AZIPROTRYN, BRASORAN, MESORANIL

2-azido-4-isopropylamino-6-methylthio-s-triazine

TYPE: Aziprotryn is a triazine compound used as a preemergence and early postemergence herbicide.

ORIGIN: 1967. ClBA-Geigy Corp.

TOXICITY: LD_{50} - 3600 mg/kg.

FORMULATION: 50% WP.

USES: Used outside the U.S. on rape, fennel, Ieek, garlic, kale, broccoli, Brussels sprouts, cabbage, onions, and others.

IMPORTANT WEEDS CONTROLLED: Pigweed, ragweed, annual bluegrass, lambsquarters, crabgrass, smartweed, foxtail, chickweed, and many others.

RATES: Applied at 1-2.5 kg ai/ha.

APPLICATIONS: Apply postemergence when weeds are between the cotyledon and 3-leaf stage. Also applied as either a layby treatment or as a preemergence application to transplant or direct-seeded crops.

PRECAUTIONS: Not for sale or use in the U.S. Cauliflower is the one crucifer crop that will be injured by the material. Toxic to fish. Do not apply after a heavy rainfall or under hot weather conditions.

ADDITIONAL INFORMATION: Soil incorporation will reduce the effectiveness. Effective through both the roots and the foliage. May be persistent in the soil for up to 120 days. Solubility in water 55 ppm.

NAMES

TERBUTHYLAZINE, GARDOPRIM, CLICK, PRIMATOL-M, TYLANEX

2-t-butylamino-4-chloro-6-ethylamino-s-triazine

TYPE: Terbuthylazine is a triazine compound used as a preemergence herbicide.

ORIGIN: 1966. CIBA-Geigy Corp.

TOXICITY: LD_{50} - 2000 mg/kg.

FORMULATIONS: 50 and 80% WP. 500 FW. Formulated with other herbicides.

IMPORTANT WEEDS CONTROLLED: A wide range of weeds and grasses.

USES: Outside the U.S. on vineyards, forests, sorghum, corn, cereals, citrus, and some fruit trees. Also as a soil sterilant.

RATES: Applied at 1.2-7.5 kg/ha.

APPLICATION: Apply prior to weed emergence. Rainfall is required to move it into the soil.

PRECAUTIONS: Moderately toxic to fish. Do not use on very sandy or gravelly soils. Not for sale or use in the U.S.

ADDITIONAL INFORMATION: Often sold formulated with other herbicides. Absorbed mainly through the root system. Slightly longer soil persistence in the soil than atrazine or simazine. Low toxicity to most fish and wildlife. Water solubility is 5 ppm.

NAMES

TERBUMETON, CARAGARD, ATHADO

2-t-butylamino-4-ethylamino-6-methoxy-s-triazine

TYPE: Terbumeton is a triazine compound used as both a pre and postemergence herbicide.

ORIGIN: 1966. CIBA-Geigy Corp.

TOXICITY: LD_{50} - 433 mg/kg.

FORMULATION: 50% WP. Formulated with other herbicides.

IMPORTANT WEEDS CONTROLLED: Annual and perennial broadleaves and grasses.

USES: Outside the U.S. in citrus, orchards, forestry and vineyards.

120

RATES: Applied at 4-10 kg actual/ha.

APPLICATION: Applied when weeds are young and actively growing.

PRECAUTIONS: Not sold in the U.S. Tree should be 3 years old before application .

ADDITIONAL INFORMATION: Often sold formulated with other compounds. Absorbed by both the leaves and the roots. Moderately toxic to fish. Rainfall is required to move it into the soil. Water solubility is 130 ppm. Long-lasting weed control .

NAMES

CYANAZINE, BELLATER, BLADEX, BLADOTLY, BLAGAL, FORTROL, PAYZE, GRAMEX, MATCH

2-(4-chloro-6-ethylamino-1,3,5-triazine-2-ylamino)2-methylpropanenitrile

TYPE: Cyanazine is a triazine compound used as a selective, preemergence herbicide.

ORIGIN: 1968. Shell Chemical company. Being sold today by DuPont.

TOXICITY: LD_{50} - 182 mg/kg.

FORMULATIONS: 4 Ib/gal flowable. 90 DF.

USES: Com, cotton, sorghum, wheat, and fallow land. Used in other countries on corn, potatoes, sugarcane, peas, ornamentals, asparagus, alfalfa, broad beans, and cereal grains. Often sold in these countries formulated with other herbicide for specific usages.

IMPORTANT WEEDS CONTROLLED: Ryegrass, barnyardgrass, crabgrass, foxtails, johnsongrass (seedlings), wild oats, witchgrass, morningglory, mustard, cocklebur, chickweed, groundsel, purslane, groundcherry, kochia, lambsquarters, smartweed, pigweed, spurge, ragweed, sherherd's purse, velvetleaf, and many others.

RATES: Applied at 1.2 - 4 Ib actual/A.

APPLICATION: Apply before planting, at planting, or after planting. If at Ieast 1/2 inch of rainfall does not fall within 4-6 days following application, a rotary hoeing or shallow cultivation is recommended. May also be applied postemergence before corn reaches the 5-leaf stage.

PRECAUTIONS: Do not apply by means of chemigation and do not apply by air. Do not use on peat or muck soils. Agitate while spraying. Do not apply to sand or sandy loam soils containing less than 1% organic matter. Do not apply postemergence when mixed with liquid fertilizers.

ADDITIONAL INFORMATION: Relatively short lived, so there should be no carry-over problems on fall-seeded crops. Active mainly through the roots, so adequate rainfall will be required to move it into the root zone. Use heavier rates on heavier soils. Solubility in water is 171 ppm. May be mixed with other herbicides to increase the weed spectrum. Provides weed control for at Ieast 10-12 weeks.

RELATED MIXTURES:

1. EXTRAZINE II—A combination of atrazine and cyanazine developed by DuPont for usage on corn.

NAMES

MEFENACET, HINOCHLOA, RANCHO

2-(2-benzothiazolyloxy)-N-methyl-N-phenylacetamide

TYPE: Mefenacet is an acetanilide compound used as a postemergence selective herbicide for grass control, especially Echinochloa spp.

ORIGIN: 1978. Bayer AG of Germany and Nihon Tokusho Nohyaku Seizo KK. of Japan.

TOXICITY: LD$_{50}$- 5000 mg/kg.

FORMULATION: 4% granules. Formulated with other herbicides.

USES: Transplanted rice outside the U.S.

IMPORTANT WEEDS CONTROLLED: Echinochloa cursgalli, Cyperus difformis, Monochoria vaginalis, Eleocharis acicularis and others.

RATES: Applied at 1.2-16 kg a.i./ha.

APPLICATION: Applied at the early growth stage of the weeds, before they reach the 3 leaf stage.

PRECAUTIONS: Crop tolerance on direct seeded rice is marginal. Not for sale or use in the U.S.

ADDITIONAL INFORMATION: This herbicide product can be applied over a longer period of time to control Echinochloa crus-galli and does not cause plant-tolerance problems in the follow-up crops.Poor control of Scirpus juncoides, Cyperus serotinus, Sagittaria pygmaea and broad-leaved weeds. Combinations with other herbicides are available.

NAMES

METAMITRON, GOLTIX, HERBRAK, GRIZZLI, COUNTDOWN

4-amino-3-methyl-6-phenyl- 1,2,4-triazin-5(4H)-one

TYPE: Metamitron is a selective triazinone herbicide with pre and postemergence activity.

ORIGIN: 1975. Bayer AG of Germany.

TOXICITY: LD$_{50}$ - 1800 mg/kg.

FORMULATION: 70% DP.

USES: Sugar beets, fodder beets and strawberries outside the U.S.

IMPORTANT WEEDS CONTROLLED: Controls many annual broadleaves and grasses, such as lambsquarters, rnatricaria, chickweed, nightshade, poa annua, and others. More effective on broadleaves than grasses.

RATES: Applied at 3.5-5 kg a.i./ha.

APPLICATION: Being used as a preplant incorporated treatment, preemergence and postemergence application. For postemergence application, additional surfactants may need to be added. Most effective when weeds are between the cotyledon and 2-leaf stage.

PRECAUTIONS: Not for sale or use in the U.S. Foxtails, crabgrass, and barnyard grass are not controlled.

ADDITIONAL INFORMATION: Shows excellent crop tolerance to sugar beets. Water solubility is 1860 ppm. Non toxic to bees or fish. Good residual activily. May be tank mixed with other herbicides. Mainly taken up by the root system .

NAMES

METRIBUZIN, **LEXONE, SENCOR, SENCORAL,**
SENCOREX, CONTRAST

4-amino-6-(1,1-dimethylethyl)-3-(methylthio)-1,2,4-triazin-5-(4H)-one

TYPE: Metribuzin is a triazine compound used as a selective pre and postemergence herbicide.

ORIGIN: 1969. Bayer AG of Gerrnany. Being sold in the U.S. by Miles Inc. and DuPont.

TOXICITY: LD_{50} - 2200 mg/kg.

124

FORMULATIONS: 70% WP. 4L. 75 DF.

IMPORTANT WEEDS CONTROLLED: Velvetleaf, ragweed, pigweed, cocklebur, jimsonweed, mustard, lambsquarters, carpetweed, sherherd's purse, smartweed, sesbania, teaweed, crabgrass, goosegrass, panicum, foxtail, and many others.

USES: Soybeans, carrots, sugarcane, alfalfa, tomatoes, corn, asparagus, chickpeas, lentils, peas, turf, wheat, barley, and potatoes. Used outside the U.S. on alfalfa, pineapple, potatoes, asparagus, soybeans, sugarcane, and tomatoes.

RATES: Used at .3-1 lb actual/A.

APPLICATION: Applied preplant incorporated, postemergenee or early postemergenee. Rain or overhead rainfall shortly after treatment is needed to activate preemergence applications. On postemergence applications, apply when weeds are less than 1.5 inches tall. To small grains, apply after crop has been fully tillered, but before jointing occurs.

PRECAUTIONS: Do not use on sandy or sandy loam soils containing Iess than 2% organic matter. Sensitive crops include crucifers, cucumbers, flax, strawberries, sugarbeets, sunflower, sweet potatoes, and tobacco. Plant soybeans seed at Ieast I 1/2 inches deep. Do not plant any other crop in treated soil but those registered for 4 months. Weeds not controlled include field bindweed, quackgrass, wild cane, groundcherry, bermudagrass, and others.

ADDITIONAL INFORMATION: Higher rates are required on soils having a high organic matter content. Water solubility 1200 ppm. Control should last for 3-4 months. May be mixed with other herbicides and fertilizers.

NAMES

ETHIOZIN, EBUZIN, TYCOR, LAKTAN, SIEGE

$$CH_3-\underset{\underset{CH_3}{|}}{\overset{\overset{CH_3}{|}}{C}}\ \begin{array}{c} O \\ \\ N-NH_2 \\ \\ S-CH_2-CH_3 \end{array}$$

4-amino-6-(1,1-dimethylethyl)-3-(ethylthio)-1,2,4-triazin-5-(4H)one (the-ethylthio analog of metribuzin)

TYPE: Ethiozin is a triazine compound used as a selective pre and postemergence herbicide.

ORIGIN: 1984. E.l Dupont de Nemours & Co. and Bayer AG of Germany.

TOXICITY: LD_{50} - 599 mg/kg. May cause skin irritation.

FORMULATIONS: 50% WP. 50 DF.

USES: Outside the U.S. on cereals and tomatoes.

IMPORTANT WEEDS CONTROLLED: Downy brome, cheatgrass, bromegrass, goatgrass, cereal rye, sunflower, foxtails, pigweed, lambsquarters, mustard, filaree, fiddleneck, henbit, chickweed, wild buckwheat, tarweed, and other annual weeds.

RATES: Applied at 0.56-1.7 kg active/ha.

APPLICATIONS: Apply when weeds are young and growing actively. Applied to cereals from emergence to the first tiller.

PRECAUTIONS: Not for sale or use in the U.S. Does not control giant ragweed, tansy mustard or bedstraw. Heavy rain soon after application may cause injury. Wheat as the following crop should be planted no less than 10 months later and other crops wait 18 months. Avoid drift. Do not treat fields that have been planted less than I " deep. Do not apply to triazine sensitive wheat varieties.

ADDITIONAL INFORMATION: May be tank mixed with other herbicides and liquid fertilizers. Rainfall is necessary to move into the root zone. Absorbed by the roots. Visual symptoms are evident in 1-3 weeks after application. Very effective on brome species.

NAMES

HEXAZINONE, GRIDBALL, VELPAR, PROMONE

3-cyclohexyl-6-(dimethylamino)-1-methyl-1,3,5-triazine-2,4(1H,3H)-dione

TYPE: Hexazinone is a triazine compound used as a selective, preemergence and postemergence herbicide.

ORIGIN: 1972. E.I. DuPont de Nemours.

TOXICITY: LD_{50}, - 1690 mg/kg. Irritating to the eyes.

FORMULATIONS: 90% SP, 2 lb/gal liquid, 75% DWG.

USES: Weed control in sugarcane, alfalfa, blueberries, pineapple, rangeland, Christmas trees, pasture grasses, and non-crop areas, such as ditch banks, storage areas, industrial sites, tank farms, railroads, highways, etc. Also being used in conifer-release programs and on forestry seedling planting sites.

IMPORTANT WEEDS CONTROLLED: Annual broadleaf and grass seedlings, as well as most annual and perennial grasses, broadleaf weeds, and vines, depending upon rates.

RATES: Applied at .2-5 lb actual/A for top kill and short-term control of established annual and perennial weeds, and 6-12 Ib/A for season-long general vegetation control.

APPLICATION: Applied postemergence during periods of active plant growth. Application when vegetation is dormant or semi-dormant may not be effective. Spray to lightly wet the foliage. Gives contact and residual control. Rainfall is required for soil activation. Apply to alfalfa that is established and dormant. For brush control, use as either a tree injection or as a basal soil treatment.

PRECAUTIONS: Do not allow roots of desired plants to come in contact with this material. Do not use near desirable trees or plants. Prevent drift. Do not apply to plants standing in water.

ADDITIONAL INFORMATION: May be used with a surfactant to improve the wetting properties. Relatively harmless to fish and wildlife. Very effective on brush and hard-to-kill perennial weeds. This material is very rate-responsive to the type of weed to be controlled and the length of control desired. More effective the higher the air temperature. Not very effective on johnsongrass. Readily absorbed by the foliage and exhibits a high degree of contact activity. Water solubility 33,000 ppm.

NAMES

CHLORSULFURON, GLEAN, TELAR, TFC

2-chloro-N-((4-methoxy-6-methyl-1,3,5-triazin-2-yl) aminocarbonyl)benzenesulfonamide

TYPE: Chlorsulfuron is a sulfonylurea compound used as a selective, preemergence and postemergence herbicide.

ORIGIN: 1979. E.l. DuPont de Nemours Co.

TOXICITY: LD_{50}- 5545 mg/kg.

FORMULATION: 75% dry-flowable granule.

USES: Wheat, barley, and oats, and non-crop weed control.

IMPORTANT WEEDS CONTROLLED: Velvetleaf, pigweed, mustards, sherherd's purse, lambsquarters, Canada thistle, sunflower, henbit, plantain, smartweed, curly dock, Russian thistle, chickweed, cocklebur, and many others.

RATES: Applied at 8-26 grams ai/ha. For residual, industrial weed control, rates of 25-150 grams ai/ha are used.

APPLICATION: Application should be made when the weeds are in the early seedling stage. If delayed beyond the 2-3-leaf stage, a higher rate may be needed. Cereals are most tolerant to postemergence treatments during the 2-leaf to tillering stage. It also can be applied to this crop preemergence. Used on fallowland to be planted back to cereals.

PRECAUTIONS: Do not use on irrigation ditches. Do not apply where treated soils may be moved by washing or blowing onto cropland. Do not apply to water saturated soils or frozen soils. Wild oats, cheat, and nightshade are tolerant of this material. Crops grown in rotation may be extremely sensitive to this material. Sugar beets, rape, and mustard are the most sensitive. Avoid drift. Do not apply next to desirable ornamentals in an industrial weed control situation. Do not store a suspension of this product for more than 2 days. Do not apply to cereals grown under flood or furrow irrigated conditions. No longer used in some areas because of weed resistance.

ADDITIONAL INFORMATION: Being used with other cereal herbicides to get wild oat control. Broadleaves are more susceptible than grasses. Shallow incorporation may improve grass control, but also may reduce broadleaf control. Plant death is often slow. Inhibits cell division in the roots and shoots. Water solubility is 125 ppm. Tolerance is obtained in crops such as cereals, by the plant degrading it to inactive products. Stops growth of weeds rapidly but it may take 1-3 weeks for them to develop noticeable symptoms. Compatible with other pesticides and liquid fertilizers. More active postemergence than preemergence. Relatively non-toxic to fish and wildlife. In non-crop weed control programs it will selectively take weeds out of many grass species. Also used to suppress grass growth and inhibit seed head formation.

NAME

FINESSE

A combination of chlorsulfuron and
metsulfuron-methyl in a 5:1 ration

TYPE: Finesse is a combination of sulfonylurea compounds used as a selective pre and postemergence herbicide.

ORIGIN: 1982. DuPont.

TOXICITY: LD_{50} - 5000 mg/kg. May cause eye irritation.

FORMULATION: 75% dry flowable.

USES: Wheat and barley.

IMPORTANT WEEDS CONTROLLED: Sowthistle, bedstraw, mustards, chickweed, buttercup, gromwell, cow cockle, fiddleneck, pennycress, groundseI, hempnettle, henbit, knotweed, kochia, lambsquarters, marestail, mayweed, smartweed, poppy, pigweed, prickly lettuce, purslane, filaree, wild buckwheat, wild radish, wild carrot, and many others.

RATES: Applied at .3-.5 oz. (75DF) per acre.

APPLICATION: Apply preplant or before plant preemergence. Rainfall or irrigation is required to move into the soil before weed seeds germinate (1/2-1 inch moisture is required.) For postemergence control apply when the crop is in the 2 leaf stage up to the boot stage.

PRECAUTIONS: Do not apply to cereals that are stressed. Do not apply preemergence to barley. Do not apply to soils with a pH greater than 7.5 or on soils with 5% or more organic matter. Persistent in the soil so it may be up to 36 months before certain crops can be rotated after the treatment. Barley should not be replanted in the treated area for 12-24 months depending upon the soil pH. Weeds that germinate after treatment and establish a root system before rainfall moves it into the soil will not be controlled. Do not apply postemergence when rainfall is threatening.

ADDITIONAL INFORMATION: Non-volatile. Both foliar and root uptake. Symptoms on the weeds will not be noticeable for 1-3 weeks following application depending upon weather conditions. When used postemergence, apply with a surfactant. May be tank mixed with other herbicides for grass control, other pesticides and with liquid fertilizer.

NAMES

METSULFURON-METHYL, **ALLIE, ALLY, BRUSHOFF, ESCORT, GROPPER, DMC**

Methyl 2-[[[[(4-methoxy-6-methyl-1,3,5-triazin-2-yl)
amino]carbonyl]amino]sulfonyl]benzoate

TYPE: Metsulfuron-methyl is sulfonylurea compound used as a selective pre and postemergence herbicide.

ORIGIN: 1982. E. I. DuPont. Sold by O.M. Scott in the turf market.

TOXICITY: LD_{50} - 5000. May cause eye and skin irritation.

FORMULATION: 60% dry flowable.

USES: Wheat and barley, and in reduced tillage fallow, non crop areas, brush control, turf and pastures.

IMPORTANT WEEDS CONTROLLED: Most broadleaf weeds.

RATES: On cereals apply at 3-8 g ai/A. On non-crop areas, use at 8-140 g/ha.

APPLICATION:

1. Cereals—Applied at the 2 leaf stage of the crop up to the boot stage.

2. Reduced tillage fallow—Apply to the fallow land and rainfall will take it into the soil.

3. Pastures—Used to control weeds in Bermudagrass pastures and turf areas. Can be used as a spot treatment on perennial weeds.

PRECAUTIONS: Long lasting, wait 22 months before planting sunflowers, flax, corn, or safflower, and 10 months before planting sorghum. Application to fescue pastures can cause stunting and seed head suppression. Do not use on ryegrass pastures or pastures containing alfalfa or clovers.

ADDITIONAL INFORMATION: Has foliar and soil activity as well as pre and postemergence activity. Non-volatile. Used as a replacement for 2,4-D. Nontoxic to fish. May be applied with other foliar herbicides. Absorbed through the Ieaves and following rainfall through the root system. It is then translocated within the plant and within hours growth is stopped.

NAMES

CHLORIMURON-ETHYL, CLASSIC

Ethyl 2-[[[[(4-chloro-6-methoxyprimidine-2-yl)amino] carbonyl] amino]sulfonyl]benzoate

TYPE: Chlorimuron-ethyl is a sulfonylurea compound used as a selective postemergence herbicide.

ORIGIN: 1982. E.I. DuPont.

TOXICITY: LD_{50} - 4102 mg/kg. May cause skin and eye irritation.

FORMULATION: 25% dispersible granule.

IMPORTANT WEEDS CONTROLLED: Cocklebur, jimsonweed, beggarweed, sesbania, morningglory, pigweed, ragweed, sicklepod, smartweed, sunflower, yellow nutsedge and others.

USES: Soybeans, peanuts and non crop areas.

RATES: Applied at .5-1 oz. of the formulation/acre.

APPLICATION: Postemergence application can be made from the 1st trifolioate soybean leaf stage has opened but no later than 60 days before maturity. For postemergence application, the weeds should be less than 4 inches tall. May be applied by air. On peanuts apply 60 days after crop emergence.

PRECAUTIONS: Do not use on soils with a pH of 7.0 or higher. Sensitive rotational crops include corn, sorghum, cotton, and rice. Consult the label for recropping interval and rotation crop guidelines.

ADDITIONAL INFORMATION: Higher rates are required on high organic matter soils. Susceptible plants will germinate and emerge in treated soils, but will then stop growing. Control takes 7-21 days. Provides partial control of some annual grasses. Weed control should last for 12 weeks after treatment. Rain 4 hours after application will not decrease the activity. May be applied in nitrogen solution. Mixed with other herbicides.

RELATED MIXTURES:

1. PREVIEW—A combination of 68.5% metribuzin and 6.5% chlorimuron-ethyl developed by DuPont to be used preemergence or preplant incorporated on soybeans.

2. LOROX PLUS—A combination of 56.9% linuron and 3.1% chlorimuron-ethyl developed by DuPont to be used preemergence on soybeans.

NAME

GEMINI

A combination of linuron and chlorimuron-ethyl in a 12:1 ratio.

TYPE: Gemini is a combination product used as a selective preemergence herbicide.

ORIGIN: 1984. DuPont & Co.

TOXICITY: LD_{50}- 1500mg/kg.

FORMULATIONS: 60% dispersible granules.

USES: Soybeans.

IMPORTANT WEEDS CONTROLLED: Cocklebur, morningglory, velvetleaf, beggerweed, lambsquarters, pigweed, purslane, ragweed, sunflower, jimsonweed, smartweed, teaweed, and others.

RATES: Applied at 12-24 oz. (60% DG) per acre.

APPLICATION: Applied as a preemergence treatment. Designed for soybeans grown in lighter soils. Rainfall or irrigation of 1/2-1 inch should follow application.

PRECAUTIONS: Do not spray over the top of emerged soybeans. Do not use on soils over pH 7. Do not use with soil applied insecticides. Consult the label for recropping intervals and crop rotation guidelines. Do not use on soils of less than 1/2% organic matter.

ADDITIONAL INFORMATION: May be applied in combination with other herbicides.

NAME

CANOPY

A combination of chlorimuron-ethyl and metribuzin in a 1:6 ratio

TYPE: CANOPY is a combination product used as a selective preemergence and preplant incorporated herbicide.

ORIGIN: 1983. DuPont Co.

TOXICITY: LD_{50} - 1500 mg/kg.

FORMULATION: 75% dispersible granule.

USES: Soybeans.

IMPORTANT WEEDS CONTROLLED: Cocklebur, jimsonweed, lambsquarters, morningglory, pigweed, teaweed, ragweed, sicklepod, smartweed, sunflower, velvetleaf, and many others. It will also suppress nutsedge.

RATES: Applied at 6-16 oz of formulation/acre.

APPLICATION: Applied either preplant incorporated or preemergence to the weeds and crop. Rainfall or irrigation is required to move it into the soil. Apply with a grass herbicide.

134

PRECAUTIONS: Grasses are not controlled. Do not use on soils with pH above 7. Avoid drift. Do not spray over the top of emerged soybeans. Do not apply to soils of less than 1/2% organic matter. Soybean injury may occur if excessive rainfall occurs after application, but before soybeans germinate. Some soybean varieties are sensitive to this product. Do not use in conjunction with soil insecticides. Consult the label for recropping intervals and rotational crop guidelines.

ADDITIONAL INFORMATION: Susceptible weeds will germinate and emerge, but growth ceases and leaves become chlorotic 3-5 days after emergence. May be mixed with other herbicides for grass control.

NAMES

TRIFENSULFURON-METHYL, HARMONY, PINNACLE, REFINE, PROSPECT

Methyl 3-[[4-methoxy-6-methyl-1,3,5-triazin-2-yl)
amino-carbonyl]amino]-sulfonyl]-2-thiophenecarboxylate

TYPE: Thifensulfuron-methyl is a sulfonylurea compound used as a selective postemergence herbicide.

ORIGIN: 1982. E. 1. DuPont de Nemours & Co.

TOXICITY: LD_{50} - 5000 mg/kg. May cause slight eye and skin irritation.

FORMULATION: 25% DG.

USES: Soybeans. Used outside the U.S. on cereals, pastures and grasslands.

IMPORTANT WEEDS CONTROLLED: Cleavers, cowcockle, pennycress, velvetleaf, smartweed, kochia, lambsquarters, tarweed, pigweed, purslane, Russian thistle, wild buckwheat, wild garlic, wild mustards, and others.

RATES: Applied at .33-.67 oz ai/A.

APPLICATION: Apply postemergence when weeds are less than 4 inches tall and before the crop canopy prevents thorough weed coverage. The crop should be form the 1st trifoliate leaf to 60 days before harvest. Surfactants will increase the activity on certain weeds.

PRECAUTIONS: Weeds not controlled include Canadian thistle, bindweed, wild oats, cheatgrass, and foxtails. Works best when weeds are actively growing. Avoid drift.

ADDITIONAL INFORMATION: Grasses are not controlled so it may be tank mixed with postemergence grass herbicides. Rapidly absorbed by the plant foliage and roots. Susceptible plants stop growing once they are sprayed and complele kill requires 7-21 days. Rotational crops can be planted in 30 days, due to rapid breakdown in the soil. May be tank mixed with other products. Works best when temperature is above 60°F. Any crop can be planted 60 days after application.

NAMES

HARMONY EXTRA, REFINE EXTRA

A combination of trifensulfron-methyl and tribenuron-methyl in a 2:1 ratio

TYPE: Harmony Extra is a combination product used as a selective postemergence herbicide.

ORIGIN: 1984. E.l. DuPont de Nemour.

TOXICITY: LD_{50} - 5000 mg/kg. Causes eye irritation.

FORMULATION: 75% DF.

USES: Wheat and barley.

IMPORTANT WEEDS CONTROLLED: Canadian thistle, chickweed, sunflower, dog fennel, henbit, kochia, lambsquarters, mustard, pigweed, Russian thistle, wild buckwheat, tarweed, wild garlic, and many others.

RATES: Applied at 26-52 grams ai/ha.

APPLICATION: Apply postemergence when the weeds are less than 4 inches tall. The crop should be in the 2 leaf boot stage but before the 3rd node is detectable.. Apply with a non-ionic surfactant.

PRECAUTIONS: Bindweed, wild oats, foxtail, and cheatgrass are not controlled. Do not tank mix with Hoelon or with malathion.

ADDITIONAL INFORMATION: Rapidly absorbed by the foliage and roots. Weeds are killed in 7-21 days. Rotational crops can be planted within 30 days of application. May be tank mixed with other herbicides and liquid fertilizers.

NAMES

TRIFLUSULFURON, DPX-66037

Methyl 2-[4-dimethylamino-6-(2,2,2-trifluoroethoxy)
-1,3,5-triazin-2-ylcarbamoylsulfamoyl]-*m*-toluate

TYPE: Triflusulfuron is a sulfonylurea compound used as a selective post-emergence herbicide.

ORIGIN: 1988, E.I. du Pont de Nemours and Company.

TOXICITY: LD_{50} 5000 mg/kg

FORMULATION: 50% dry flowable.

USES: Experimentally on sugar beets.

IMPORTANT WEEDS CONTROLLED: Annual smartweed, annual sowthistle, bed-straw, common lambsquarters, common ragweed, foxtail species, nettleleaf goosefoot, hairy nightshade, knotweed, kochia, pigweed species, Russian thistle, velvetleaf, wild buckwheat, wild mustard, wild radish, common mallow and many others.

RATES: Applied at 15-35 g a.i./ha.

APPLICATION: Apply to the weeds when they are in the cotyledon to 4-leaf stage and

137

actively growing. Sequential applications 7-14 days apart enhance the activity. Use with a non-ionic surfactant or crop oil concentrate.

PRECAUTIONS: Used on an experimental basis only. Avoid drift.

ADDITIONAL INFORMATION: May be tank-mixed with other postemergence herbicides. Most effective on broadleaf weeds. Short-lived in the soil so other crops can be planted within 60 days of application.

NAMES

RIMSULFURON, DPX-E 9636, TITUS

N-[[(4,6-dimethoxy-2-pyrimidinyl)amino]carbonyl]-=3 (ethylsulfonyl)-
2-pyridinesulfonamide

TYPE: Rimsulfuron is a sulfonylurea compound used as a selective postemergence herbicide.

ORIGIN: DuPont, 1989.

TOXICITY: LD_{50} 5000 mg/kg. May cause eye irritation.

FORMULATION: 25% DF.

USES: Experimentally being used on corn.

IMPORTANT WEEDS CONTROLLED: Quackgrass, wild oats, crabgrass, barnyardgrass, foxtails, johnsongrass, panicum, pigweed, bedstraw, kochia, mustard, chickweed and many others.

RATES: Applied at 8-35 g a.i./ha.

APPLICATION: Apply when grasses are in the 2-5 leaf stage and broadleaves are in the

2-6 leaf stage. Use with a non-ionic surfactant or crop oil concentrate. A split application may be required to control perennial grasses.

PRECAUTIONS: Used on an experimental basis only. Selectively may be decreased if organophosphate insecticides are applied to the soil or as a foliar treatment. Do not use on sweet corn. Do not use on corn that is over the 7 leaf stage. Do not apply under hot weather conditions (over 25 C) at the time of application.

ADDITIONAL INFORMATION: Controls some broadleaf weeds that are triazine resistant. Rapidly metabolized by the corn plant. Treated weeds stop growing soon after treatment. May be tank mixed with other herbicides and liquid fertilizer. Very little soil residual activity.

NAMES

TRIBENURON-METHYL, **EXPRESS, GRANSTAR, CAMEO, POINTER**

Methyl 2-1[[[[n-3-(4-methoxy-6-methyl-1,3,5-triazin-2-yl)methylamino]carbonyl]amino]sulfonyl]benzoate.

TYPE: Tribenuron-methyl is a sulfonylurea compound used as a selective postemergence herbicide.

ORIGIN: 1984. E.l. DuPont de Nemours & Co.

TOXICITY: LD_{50}, - 11,000 mg/kg. May cause skin and eye irritation.

FORMULATION: 75% DF.

USES: Barley and wheat.

IMPORTANT WEEDS CONTROLLED: Fiddleneck, Canadian thistle, kochia, sunflower, lambsquarters, mustard, henbit, Russian thistle, and many other broadleaf weeds.

RATES: Applied at 9-35 grams ai/ha.

APPLICATION: Apply postemergence when weeds are young and growing actively, usually Iess than 4 inches tall. Use with a non-ionic surfactant. Crop should be from the 2 Ieaf to before the flag leaf is visible.

PRECAUTIONS: Weeds not controlled include wild buckwheat, bindweed, wild garlic, wild oats, foxtail, and cheatgrass. Rotational crops can be planted in 60 days.

ADDITIONAL INFORMATION: Non-cereal crops can be planted as soon as 60 days after treatment. May be tank mixed with other herbicides. Readily absorbed and translocated throughout the plant. Weeds are killed in 7-21 days. May be used in liquid fertilizers.

NAMES

BENSULFURON-METHYL, LONDAX, MARINER, ROZAL, WE-HOPE

Methyl 2[[[[[(4,6-dimethoxyprimidin-2-yl) amino]
carbonyl] amino]sulfonyl]methyl]benzoate

TYPE: Bensulfuron-methyl is a sulfonylurea compound used as a selective pre and postemergence herbicide.

ORIGIN: 1982. DuPont Co.

TOXICITY: LD_{50} - 5000 mg/kg. May cause eye irritation.

FORMULATIONS: 60% DF. Formulations used other herbicides for usage outside the U.S.

USES: Rice. Experimentally being used as an aquatic herbicide.

IMPORTANT WEEDS CONTROLLED: Monochoria,arrowhead, waterplantain, umbrellasedge, hydrilla, bulrush, and others. Barnyardgrass is moderately susceptible.

RATES: Applied at 20-75 g a.i./ha.

APPLICATION: Applied either pre or early postemergence as a broadcast treatment when the rice is in the 1-3 leaf stage.. The water should be neither flushed or drained for 7 days after application . Application can be made up to the 3 leaf stage of the weeds.

PRECAUTIONS: Not effective on sprangletop. Draining the field after application will result in poor weed control. Result may be delayed if air or water temperatures are below 65°F. Product performance will be reduced in areas not covered with water. Do not use on wild rice. Do not use water drained directly from treated fields to irrigate other crops. Do not farm crawfish in treated fields.

ADDITIONAL INFORMATION: Best results are obtained by early postemergence application. It is most effective on annual and perennial broadleaf weeds and sedges. Indica rice varieties are more tolerant to this product than the Japonica type rice varieties. Safe on fish. The higher the temperature, the more rapid the activity on the weeds. After application into the water the active ingredient is rapidly released and subsequently absorbed by the plant. Results can be seen in 3-5 days and complete control in 7-21 days. May be used with other rice herbicides.

SULFOMETURON-METHYL, OUST

Methyl 2-[[[[(4,6-dimethyl-2-pyrimidinyl) amino]-carbonyl] amino]sulfonyl]benzoate

TYPE: Sulfometuron-methyl is a sulfonylurea compound used as a broad spectrum, pre and postemergence herbicide.

ORIGIN: 1980. E.l. DuPont de Nemours & Co.

TOXICITY: LD_{50} - 5000 mg/kg. May cause slight eye irritation.

FORMULATION: 75% dry-flowable granule.

IMPORTANT WEEDS CONTROLLED: Johnsongrass, nutsedge, Canadian thistle, curly dock, kudzu, poison ivy, turkey mullein, wild blackberries, plantain, dandelion, horsetail, kochia, mustards, pigweed, ragweed, Russian thistle, sunflower, annual bluegrass, ryegrass, barnyard grass, foxtails, bromes, canarygrass, and many others.

USES: Used for non-selective weed control in non-crop areas. For selective weed control, it can be used on non-cropland to control weeds where bermudagrass exists. Also used in forestry on pine trees preplant or in a conifer release program. May be used in the selective weed control program for the release of bermudagrass, bahiagrass, smooth brome and crested wheatgrass.

RATES: Applied at 1-12 oz (75 DF) per acre.

APPLICATION: Apply preemergence or early postemergence for annual weed control. On johnsongrass, apply postemergence when it is actively growing. A second application may be necessary if regrowth occurs. May be used with a surfactant to increase the degree of control. May be applied by air in forestry.

PRECAUTIONS: Avoid drift. Up to .5 - I inch of rainfall is needed to take the product into the soil . Do not apply to frozen ground. Do not apply near desirable trees and plants. Tolerant weeds include bermudagrass, buffalograss, milkweed, bindweed, dallisgrass, nightshade, groundcherry, horsenettle, and others. Incompatible with alkaline pesticides. Do not store the spray mixture more than two days. Do not use where the roots of desirable plants are growing. Do not apply to ditches, ponds, or reservoirs that are used for irrigation water. Do not apply where treated soils can be moved by washing or blowing onto cropland. Do not apply to water saturated soils. Injury may occur if used on Douglas fir or Ponderosa pine.

ADDITIONAL INFORMATION: Does not move laterally in the soil. Not broken down by sunlight. Absorbed through both the roots and leaves. May take 4-6 weeks for complete kill. Rainfall is required to activate the material. This product does not inhibit seed germination. The halflife in the soil under summer field conditions is about 4 weeks. May be applied in combination with other herbicides.

NAMES

NICOSULFURON, ACCENT, NISSHIN, DPX-V9360, SL-950, CHALLENGER

2-[[[(4-6-dimethoxypyrimidin-2-yl)-aminocarbonyl] aminosulfonyl]-N,N-dimethyl-3-pyridinecarboximide

TYPE: Nicosulfuron is a sulfonylurea compound used as a selective postemergence herbicide.

ORIGIN: 1986. ISK of Japan. Being marketed in the U.S. by DuPont.

TOXICITY: LD_{50} - 5,000 mg/kg. May cause slight eye irritation.

FORMULATION: 75% DG, 40 SC.

USES: Corn.

IMPORTANT WEEDS CONTROLLED: Johnsongrass, quackgrass, foxtails, shattercane, panicums, barnyardgrass, sandbur, pigweed, morningglory and others.

RATES: Applied at 50-100 g ai/ha.

APPLICATION: Applied postemergence with a non-ionic surfactant when weeds are 4-12 inches tall and actively growing. Corn should be in the 2-6 leaf stage. Repeat applications may be necessary.

PRECAUTIONS: Sorghum is very sensitive to this product. Breaks down in the soil slower under alkaline than acidic soil conditions. Delay the use of this product for at least 7 days after the use of an organic phosphate foliar insecticide. Do not apply to corn taller than 35 inches. Do not apply to corn that has been previously treated with Counter. Do not apply to corn that has been treated within 7 days before with 2,4-D, Basagran or Laddok. Temperatures below 50°F will result in poorer control.

ADDITIONAL INFORMATION: May be tank mixed with other compounds. Corn is very tolerant to the product. Treated weeds stop growing after treatment and symptoms will appear in a few days with complete death in 1-3 weeks. Rain within 2 hours of application will not decrease the effectiveness.

ETHAMETSULFURON-METHYL, DPX-A 7881, MUSTER

Methyl 2-[(4-ethoxy-6-methylamino-1,3,5-triazin-2-yl)carbamoylsulphamoyl] benzoate

TYPE: Ethametsulfuron-methyl is a sulfonylurea compound used as a selective postemergence herbicide.

ORIGIN: 1984. DuPont Co.

TOXICITY: LD_{50} - 5,000 mg/kg. May cause eye irritation.

FORMULATION: 75% DF.

USES: Outside the U.S. on rape.

IMPORTANT WEEDS CONTROLLED: Wild mustard, stinkweed, smartweed, clovers, hempnettle, pigweed and others.

RATES: Applied at 10-120 g a.i./ha.

APPLICATION: Apply to the crop after it has reached the 2-leaf stage but before the canopy covers the weeds. Weeds should be young and actively growing. Apply with a recommended surfactant.

PRECAUTIONS: Avoid drift. Apply only by ground equipment. Do not contaminate irrigation water. Used on an experimental basis only in the U.S.

ADDITIONAL INFORMATION: May be tank mixed with grass herbicides such as Poast. Rape shows a 4 fold safety margin to this product. Absorbed by both the foliage and the roots. Slow acting. Higher rates are required when used on winter rape than on the spring crop.

NAMES

TRIASULFURON, CGA 131036, LOGRAN, AMBER, KEOS

2-(2-chloroethoxy)-n-[[(4-methoxy-6-methyl-1,3,5-
triazin-2-yl) amino]carbonyl] benzene sulfonamide

TYPE: Triasulfuron is a sulfonylurea compound used as a selective pre and postemergence herbicide.

ORIGIN: 1985. CIBA-Geigy Ltd.

TOXICITY: LD_{50} - 5000 mg/kg. May cause slight eye and skin irritation.

FORMULATION: 75% water dispersible granules.

USES: Wheat and barley and fallowland.

IMPORTANT WEEDS CONTROLLED: Charlock, chickweed, common poppy, corn marigold, field pansy, field pennycress, hempnettle, kochia, mayweed, pigweed, shepherd's purse, smartweed, wild buckwheat, wild radish and many others.

RATES: Applied at 5-30 g. ai/ha.

APPLICATION: Application should be made when the weeds are in the early seedling stage, and when the weeds are actively growing. Application in the cereal crop (winter and spring) should be during the 2-leaf to booting stage for postemergence applications. Preemergence applications can be made on the fallow land in fallow cereal rotational systems. For postemergence applications, the use of a surfactant in the spray mixture is recommended to improve wetting and contact activity of the product. Based on the broadleaved spectrum of activity, the product provides excellent complementary activity for substituted urea herbicides and specific mixtures have been developed with chlortoluron and isoproturon.

PRECAUTIONS: Avoid drift. Do not apply if rainfall is expected within 24 hours. Do

146

not use within 60 days of an organic phosphate insecticide application. Crops other than wheat and barley can be very sensitive to low concentrations of this product in the soil.

ADDITIONAL INFORMATION: Based on preemergence applications (wheat only) the product provides grass control (Lolium spp.) and the activity is enhanced by shallow incorporation. Growth of susceptible weeds is very rapidly inhibited immediately after application. Further symptoms become visible 1-3 weeks after application. Less susceptible weeds may only be inhibited and suppressed without dying completely. May be tank mixed with other herbicides. Absorbed through both roots with foliage. May be applied by air and mixed with liquid fertilizer.

NAMES

PRIMISULFURON-METHYL, BEACON, CGA-136872, RIFLE, TELL

3-[4,6-bis(difluoromethoxy)pyrimidin-2-yl]-1-(2-methoxycarbomyl-
phenylsulfonyl)-urea

TYPE: Primisulfuron-methyl is a sulfonylurea compound used as a selective, postemergence herbicides.

ORIGIN: 1985. CIBA Geigy.

TOXICITY: LD_{50} - 5050 mg/kg. May cause slight eye irritation.

FORMULATIONS: 75% WG

USES: Corn, non crop areas and turf.

IMPORTANT WEEDS CONTROLLED: Beggerweed, cocklebur, jimsonweed, mustard, pigweed, ragweed, johnsongrass, shattercane, prickly sida, sunflower, nightshade, sesbania, kochia, quackgrass, velvetleaf and others.

147

RATES: Applied at 20-40 g. ai/ha

APPLICATION: Applied postemergence when corn is in the 3-7 leaf stage (4-20 inches tall). Use with a spray adjuvant to improve the control. The weeds should be no more than 9 inches in height or diameter. Apply over the top, directed or semi-directed spray by ground only.

PRECAUTIONS: Corn hybrids differ in sensitivity to this herbicide. Rainfall within 4 hours will reduce the effectiveness. Do not apply to corn when the insecticide terbufos has been previously applied. Do not apply foliar organo-phosphate insecticides 10 days before or after an application of this product.

ADDITIONAL INFORMATION: Compatible with other herbicides. Taken up by both the foliage and roots of the plant. Susceptible plants appear yellow in 3-5 days followed by complete control in 7-21 days. Half life in the soil is 3-10 weeks. Also provides residual preemergence control. May be used with liquid fertilizers.

NAMES

HALOSULFURON-METHYL, NC 319, MON-12000, PERMIT, BATTALION

Methyl-3-chloro-5-(4,6-dimethoxypyrimidin-2-ylcarbamoylsufamoyl)
-1-methylpyrazole-4-carboxylate

TYPE: Halosulfuron-methyl is a sulfonylurea compound used as a selective pre and postemergence herbicide.

ORIGIN: Nissan Chemical Co. of Japan 1988. Being developed in the U.S. and certain other countries by Monsanto.

TOXICITY: LD_{50} 8865 mg/kg. May cause slight eye irritation.

FORMULATION: 50% WP, 75% DF.

USES: Experimentally being tested on corn, grain sorghum, rice, sugarcane and turf.

IMPORTANT WEEDS CONTROLLED: Cocklebur, sandbur, kochia, lambsquarters, nutsedge, pigweed, ragweed, sicklepod, smartweed, velvetleaf, mallow and other broadleaf weeds.

RATES: Applied at 18-140 g a.i./ha.

APPLICATION: On corn applied preplant incorporated, preemergence, and early postemergence. If used preemergence on corn a safener (MON-13900 must be used). Can be used as a lay-by treatment. Apply postemergence when weeds are under 12 inches tall. When used postemergence, use with a non-ionic surfactant or crop oil concentrate.

PRECAUTION: Used on an experimental basis only. Grasses are not controlled.

ADDITIONAL INFORMATION: May be used in combination with other herbicides. Gives season long control. Controls nutsedge, postemergence.

NAMES

FLAZASULFURON, SHIBAGEN, SL-160

1-(4,6-dimethoxypyrimidin-2-yl)-3-(3-trifluoromethyl-2-pyridysulphonyl) urea

TYPE: Flazasulfron is a sulfonyl-urea compound used as a selective postemergence herbicide.

ORIGIN: Ishihara Sangyo Kaisha Ltd. of Japan 1988.

TOXICITY: LD_{50} 5000 mg/kg. May cause eye irritation.

FORMULATION: 10% WP.

IMPORTANT WEEDS CONTROLLED: Foxtails, crabgrass, barnyardgrass, sedges, pigweed, annual bluegrass, oxalis, shepardspurse, chickweed and others.

USES: Being developed on bermudagrass and zoysiagrass turf.

RATES: Applied at 25-100 g a.i./ha.

APPLICATION: Applied early postemergence with the weeds are in the 2-4 leaf stage. May be applied at anytime of the year.

PRECAUTION: Not for sale or use in the U.S. Do not use on cool season grasses. Under adverse conditions a yellowing of the turf will occur but it is quickly outgrown. Rain within 24 hours will decrease its effectiveness.

ADDITIONAL INFORMATION: Has both soil and foliar activity but most active through the foliage. Residual soil activity is 1-3 months. Controls both annual and perennial weeds. Very effective on Cyperus species. Weeds stop growing immediately after treatment and discoloration will be noted in 4-5 days with 20-30 days for complete kill.

NAMES

IMAZOSULFURON, TH-913, TAKEOFF, SIBATITO

1-(2-chloroimidazo [1,2-*a*] pyridin-3-ylsulfonyl)-3-(4,6 dimethoxyprimidin 2-yl) urea

TYPE: Imazosulfuron is a sulfonyl-urea compound used as a selective preemergence and early postemergence herbicide.

ORIGIN: Takeda Chemical Int. 1989.

TOXICITY: LD$_{50}$ 5000 mg/kg.

FORMULATION: Experimentally on paddy rice and on turf.

IMPORTANT WEEDS CONTROLLED: Many species of annual and perennial broadleaf weeds and sedges.

RATES: Applied at 50-100 g a.i./ha.

APPLICATION: Applied as a preemergence treatment.

PRECAUTIONS: Not for usage in the U.S. Used on an experimental basis only.

ADDITIONAL INFORMATION: Registration in Japan is expected in 1993.

NAMES

AMIDOSULFURON, **GRATIL, ADRET, HOE 075032**

1-(4,6-dimethoxypyrimidin-2-yl)-3-(mesyl) sulfamoylurea

TYPE: Amidosulfuron is a sulphonylurea compound used as a selective postemergence herbicide.

ORIGIN: Hoechst Ag of Germany 1989.

TOXICITY: LD_{50} 5000 mg/kg.

FORMULATION: 20 WDG, 75 WDG

USES: Experimentally being tested on cereal, rice, flax and potatoes.

IMPORTANT WEEDS CONTROLLED: Cleavers, shepardspurse, mustards, dock, smartweed, wild radish, wild buckwheat, chickweed, sweetclover, pigweed, bindweed and many other broadleaf weeds.

RATES: Applied at 15-90 g a.i./ha.

APPLICATION: Apply postemergence to the weeds when the cereals are from the 2-4 leaf stage, to the flag stage.

PRECAUTIONS: Not for sale or use in the U.S.

ADDITIONAL INFORMATION: Low temperature at the time of application will slow down the activity but will not unfavorably affect the degree of control. Rain that occurs 1 hour after application will not reduce the activity. May be mixed with other herbicides. All cereals appear to be tolerant. Taken up by the roots and leaves and translocated within the plant. It requires 2-4 weeks for complete control. The half life in the soil is 14-29 days.

NAMES

CINOSULFURON, SETOFF, CGA-142464

3-(4,6-dimethoxy-1,3,5-triazin-2-yl)-1-[2-(2-methoxyethoxy)phenylsulfonyl]-urea

TYPE: Cinosulfuron is a sulfonylurea herbicide used pre and postemergence in rice.

ORIGIN: 1985. CIBA-Geigy Ltd.

TOXICITY: LD_{50} 5000 mg/kg.

FORMULATION: 20% WDG. Formulated with other herbicides.

USES: Outside the U.S. on transplanted rice, pre-germinated wet sown rice, water sown rice, and dry sown flooded rice. Also used on plantation crops.

IMPORTANT WEEDS CONTROLLED: Waterplantain, umbrella plant, yellow and other sedges, marsilea, pickerel weed, pondweed, toothcup, arrowhead, balrush, gooseweed and many others.

RATES: Applied at 10-80 g a.i./ha.

APPLICATION: The low rates are applied on irrigated rice in tropical countries whereas hlgher rates are needed in temperate climates and on dryland rice. Granular formulations are either applied by hand or with granule spreaders. Also used as a foliar spray as a postemergence treatment.

PRECAUTIONS: Not for sale or use in the U.S. Tolerance of Japonica type rice species may be marginal for preemergence usage.

ADDITIONAL INFORMATION: Barnyardgrass is not controlled so usually combined with other herbicides.

NAMES

IMAZAMETHABENZ-METHYL, AC 222,293, ASSERT, DAGGER, MEGAPLUS

2-[4,5-dihydro-4-methyl-4-(1-methylethyl)-5-oxo-1H-imidazol-2-yl) -4(and 5)-methylbenzoic acid (3:2)

TYPE: Imazamethabenz-methyl is a mixture of two positioned isomers that are imidazole compounds used as selective, postemergence herbicides.

ORIGIN: 1982. American Cyanamid Company.

TOXICITY: LD_{50} - 2333 mg/kg. May cause skin and eye irritation.

FORMULATION: 2.5 LC. Formulated with other herbicides in some countries.

USES: Wheat, barley and sunflower. Being used on small grains in Europe.

153

IMPORTANT WEEDS CONTROLLED: Wild oats, blackgrass, silky bentgrass, wild mustard, wild buckwheat, field pennycrest, wild radish and others.

RATES: Applied at .18-.461b ai/acre in combination with a non-ionic surfactant.

APPLICATION: Applied when the wild oats are in the 1-4 leaf stage by air or ground. Small grains should be past the 2 leaf stage but before development of the first internode.

PRECAUTIONS: A non-ionic surfactant is essential for optimum performance. Ryegrass, bromegrass, foxtails and canarygrass are not controlled. Do not plant oats, rape, mustard, broccoli or lentils for 15 months after application. Do not plant sugarbeets for 20 months. Do not tank mix with Banvel, 2,4-D amine, or MCPA amine.

ADDITIONAL INFORMATION: Suppresses bedstraw, pigweed, kochia, Russian thistle, and other broadleaves. Can be combined with other herbicides. Absorbed through the foliage and roots and translocated to the meristematic regions. Slow-acting, although growth is stopped, death to the weed may not occur for several weeks. May be applied by air. May be applied with liquid fertilizers.

NAMES

IMAZAQUIN, IMAGE, SCEPTER, STYMIE, TONE-UP

2-[4,5-dihydro-4-methyl-4-(1-methylethyl)-S-oxo-1,H-imidazol-2-yl]-3 quinolinecarboxylic acid

TYPE: Imazaquin is a imidazole compound used as a selective, pre and postemergence herbicide.

ORIGIN: 1981. American Cyanamid Company.

TOXICITY: LD_{50} - 5000 mg/kg. May cause skin irritation.

FORMULATION: 1.5 EC. 70DG. Formulated with other herbicides.

USES Soybeans, turf and ornamentals.

IMPORTANT WEEDS CONTROLLED: Cocklebur, pigweed, prickly sida, nightshade, mustard, jimsonweed, morningglory, smartweed, lambsquarters ragweed, velvetleaf, foxtail, and others.

RATES: Applied at .125-.375 kg ai/ha.

APPLICATION: Found to be effective when used preemergence, postemergence, or preplant incorporated, 1-2 inches deep. When used postemergence apply before the weeds are 12 inches high. Use with a surfactant. On turf applied either preemergence or postemergence to the weeds. Water with 1-7 days to wash it into the rootzone.

PRECAUTIONS: Velvetleaf must be treated postemergence before the 2-leaf stage. Temporary yellowing to turf may occur. Avoid drift. Do not use on turf greens. Do not tank mix with postemergence grass herbicides. Corrosive to steel and aluminum. Do not use on container grown ornamentals. Do not use on turf during periods of slow growth. Severe injury has been noted on the following ornamentals; Azalea, Viburnum, Pieris, Abelia and Ligustrum. Persists in the soil.

ADDITIONAL INFORMATION: Soybeans are especially tolerant to this product and the tolerance increases with age, so late postemergence applications are a possibility. Broad leaves are more sensitive than grasses. Absorbed by both the roots and foliage and then translocated. Very fast-acting on cocklebur and pigweed, but with other susceptible weeds the activity is relatively slow. Used postemergence with a non-ionic surfactant. Inhibits a key enzyme that is necessary to produce a vital amino acid, thereby controlling the weed. May be applied by air. It may be used preemergence on postemergence on bermudagrass, centipede grass, St. Augustine grass or zoysiagrass turf for annual weed control. May be tank mixed with other products.

RELATED MIXTURES:

I . SQUADRON—A combination of Imazaquin and pendimethalin developed by American Cyanamid to use on soybeans either preplant incorporated or preemergence.

2. TRI SCEPT—A combination of Imazaquin and trifluralin to be used preplant incorporated on soybeans. Developed by American Cyanamid.

3. SCEPTER O.T.—A combination of Imazaquin and acifluorfen-sodium developed by American Cyanamid for use on soybeans.

IMAZETHAPYR, PIVOT, PURSUIT, HAMMER, OVERTOP

CH₃—CH₂ ... C—OH ... N ... CH₃ CH₃ ... CH ... HN ... CH₃ ... O

2-[4.5-dihydro-4-methyl-4-(1 -methylethyl)-5-oxo-lH-imidazol-2-yl]-5-ethyl-3-pyridinecarboxylic acid

TYPE: lmazethapyr is an imidazole compound used as a selective pre and postemergence herbicide.

ORIGIN: 1985. American Cyanamid Co.

TOXICITY: LD_{50} - 5,000 mg/kg. May cause eye and skin irritation.

FORMULATION: 2 Ib/gal. aqueous solution. Formulated with other herbicides.

USES: Soybeans, peanuts and legume vegetables. Experimentally being tested on alfalfa, clover and other crops. Being sold on these outside the U.S.

IMPORTANT WEEDS CONTROLLED: Barnyardgrass, crabgrass, foxtails, panicums, jimsonweed, lambsquarters, nightshade, pigweed, velvetleaf, cocklebur, smartweed, momingglory, mustards, seedling johnsongrass, and others.

RATES: Applied at .032 - .125 Ib. ai/A.

APPLICATIONS: Applied as a preplant incorporated, preemergence and postemergence application. If used postemergence, combine with a surfactant when weeds are actively growing, but less than 3 inches tall.

PRECAUTIONS: Sicklepod and sesbania are not controlled. Do not store below 32°F. Corrosive to steel and aluminum.

ADDITIONAL INFORMATION: May be combined with other herbicides. Absorbed by the roots and foliage of the plant and translocated to the meristematic regions. When applied postemergence, weeds stop growing soon after treatment and die within 2-4 weeks. May be used with liquid fertilizers.

RELATED MIXTURES:

1. PURSUIT PLUS— A combination of Imazethapyr and pendimethalin to be used preplant incorporated on soybeans. Developed by American Cyanamid.

2. PASSPORT—A combination of Imazethapyr and trifluralin developed by American Cyanamid to be used as a preplant incorporated treatment on soybeans.

NAMES

IMAZAPYR, ARSENAL, ASSAULT, CHOPPER, CONTAIN

2-[4,5-dihydro-4-methyl-4-(1-methylethyl)-5-oxo-1H-imidazol-2-yl]-3-pyridinecarboxylic acid

TYPE: Imazapyr is an imidazole compound used as a non-selective pre and postemergence herbicide.

ORIGIN: 1981. American Cyanamid.

TOXICITY: LD_{50} - 5000 mg/kg. May cause eye and skin irritation.

FORMULATION: 2EC, 5% granules.

USES: Being used as a broad spectrum herbicide in non-crop areas and forestry. Used outside the U.S. on rubber and coconut plantations.

IMPORTANT WEEDS CONTROLLED: Crabgrass, barnyardgrass, foxtails, velvetleaf, pigweed, mustards, lambsquarters, spurge, morningglory, purslane, quackgrass, bermudagrass, bindweed, dock, brush species, vines, brambles and many others.

RATES: Applied at .4-2 kg ai/ha.

APPLICATION: Can be applied either preemergence or postemergence to the weeds. For postemergence applications, use with a non-ionic surfactant and apply when weeds are growing vigorously. Split applications may be required on some perennial weeds. In forestry it can be painted on cut strips to prevent regrowth, to use in conifer release programs, in site preparation and for use to control weeds in conifer nurseries. May be injected into hatchet cuts around the tree, as a basal bark treatment and as a cut strip treatment. Apply to woody species in the late summer or fall or in the winter months. Controls undesirable weeds in unimporoved bermudagrass and bahiagrass area. May be used under asphalt. May be applied by air.

PRECAUTIONS: Avoid runoff or application near desired plants. Do not store below 10°F.

ADDITIONAL INFORMATION: A broad spectrum herbicide. Absorbed through the foliage and roots. After application the plant quits growing, but complete control may take several weeks. Translocated into the root system. Woody species are also controlled, but require a higher rale. Persisls for 3 months to I year in the soil, depending upon dosage and soil moisture. Lateral and vertical movement in the soil is limited. Conifers are tolerant to this product. Used in conifer release programs. When applied to woody species during the dormant period leaf out in the spring is often prevented.

UREA COMPOUNDS

NAMES

DIURON, DCMU, DIATER, DAILON, VONDURON, DIREX, DIUREX, DMU, KARMEX, MARMER, UNIDRON

3-(3,4-dichlorophenyl)-1,1-dimethylurea

TYPE: Diuron is a substituted urea compound used as a pre and postemergence herbicide.

ORIGIN: 1951. E.l. DuPont de Nemours Chemical Company. Produced today by a number of manufacturers.

TOXICITY: LD_{50} - 3400 mg/kg. May cause irritation to the eyes and skin.

FORMULATIONS: 4L, 80% WP, 80 DF. Formulated with other herbicides.

USES: Cotton, apples, caneberries, oats, pecans, papaya, gladiolus, trefoil, barley, pears, bermudagrass, corn, sugarcane, pineapple, peppermint, walnuts, sorghum, alfalfa, macadamia nuts, blueberries, bananas, perennial grasses, olives, grapes, asparagus, gooseberries, artichokes, wheat, citrus, and non-crop areas for total vegetation control.

IMPORTANT WEEDS CONTROLLED: Crabgrass, barnyardgrass, foxtail, johnsongrass, pigweed, purslane, ragweed, chickweed, mustard, ryegrass, annual morningglory, lambsquarters, and many others.

RATES: Applied at .5-20 lb a.i./ha. The higher rates are used for total vegetation control.

APPLICATION: Applied as a preemergence treatment prior to weed germination. Rainfall is required to take it into the soil. Used as a lay-by treatment on some crops. Often used with other herbicides.

PRECAUTIONS: Do not apply near desired plants. Do not use on turf. On most applications, depending on the rate used, do not plant a susceptible crop within 12 months of application. Do not treat dwarf varieties of fruit trees.

ADDITIONAL INFORMATION: This chemical stays near the soil surface due to its ability to resist leaching, since it has a low water solubility and is absorbed by the soil colloids. A non-corrosive, non-volatile compound. Does not leach over one inch when applied at proper rates. It may not provide satisfactory control of hard-to-kill, deep-rooted perennial weeds. Preferred over monuron as a soil sterilant in areas of high rainfall and/or light, sandy soils. Does not persist as long in the soil as monuron. Less soluble in water than monuron and fenuron, but more soluble than neburon. More strongly absorbed by the soil so it leaches more slowly than monuron. The use of a surfactant greatly increases its knockdown effects or contact activity. May be applied by air.

NAMES

NEBURON, NEBUREA, NEBUREX, NORUBEN

1-N-butyl-3-(3,4-dichlorophenyl)-1-methylurea

TYPE: Neburon is a substituted urea used as a selective, preemergence herbicide.

ORIGIN: 1955. E.I. DuPont de Nemours Chemical Company. Sold outside the U.S. by a number of companies.

TOXICITY: LD_{50} - 11,000 mg/kg. May be irritating to the eyes and skin.

FORMULATION: 60% WP. Formulated with other herbicides.

USES: Used outside the U.S. on wheat, strawberries, alfalfa, and nursery plantings.

IMPORTANT WEEDS CONTROLLED: Chickweed, lambsquarters, ragweed, smartweed, purslane, pigweed, crabgrass, annual bluegrass, and others.

RATES: Applied at 2-3 kg a.i./ha.

APPLICATION: Treat just prior to the germination and growth of annual weeds. Best results are obtained when it is moved into the rootzone by moisture within two weeks of application. Do not disturb the surface of the soil after treatment.

PRECAUTIONS: Persists in the soil. Avoid spraying foliage, or injury to some desirable plant species may result. Not for sale or use in the U.S.

ADDITIONAL INFORMATION: Slow acting. Non-corrosive and non-volatile. Resists leaching. Less water soluble than diuron and monuron.

NAMES

SIDURON, TUPERSAN

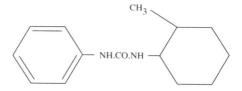

[1-(2-methylcyclohexyl)-3-phenylurea]

TYPE: Siduron is a substituted urea compound used on a selective preemergence herbicide.

ORIGIN: DuPont and Co. 1964.

TOXICITY: LD_{50} 7500 mg/kg.

FORMULATION: 50% WP.

USES: Turf.

IMPORTANT WEEDS CONTROLLED: Crabgrass, foxtails, barnyardgrass and others.

RATES: Apply at 2-6 lbs a.i./acre.

APPLICAITON: On new plantings apply as the final operation following seeding. A second application may be applied 30 days later. On established turf, apply in the spring before expected weed emergence. May be applied at any stage of turf development.

PRECAUTION: Do not use on Bermudagrass turf or certain strains of bentgrasses. Do not use on golf greens. Poa annual, clover and broadleaf weeds are not controlled.

ADDITIONAL INFORMATION: For usage on bluegrass, fescue, redtop, smoothbrome, perennial ryegrass, orchardgrass, zoysia and certain stages of bentgrass turf. At least .5 inch of water by irrigation or rainfall must be applied within 3 days of application. May be mixed with fertilizers and other herbicides.

NAMES

CHLOROXURON, CHLOROXIFENIDIM, NOREX, TENORAN

Cl—⟨benzene ring⟩—O—⟨benzene ring⟩—N(H)—C(=O)—N(CH₃)(CH₃)

N'-[4-(4-chlorophenoxy) phenyl]-N',N'-dimethylurea

TYPE: Chloroxuron is a substituted urea and is used as a selective, preemergence and postemergence herbicide.

ORIGIN: 1961. CIBA-Geigy Corp. No longer used or produced in Ihe U.S.

TOXICITY: LD_{50} - 3000 mg/kg. May cause slight eye and skin irritation.

FORMULATION: 50% WP.

USES: Outside the U.S. on carrots, onions, soybeans, celery, strawberries, Ieeks, turf, conifers, and ornamentals.

IMPORTANT WEEDS CONTROLLED: Cocklebur, morningglory, pigweed lambsquarters, ragweed, Florida pusley, puncture vine, crabgrass, goosegrass, barnyardgrass, nightshade, carpetweed, chickweed, groundsel, jimsonweed, purslane, shepherd's purse, spurry, velvetleaf, mustard, and others.

RATES: Applied al 1-4.5 kg ai/ha.

APPLICATION: Used as a preemergence or postemergence treatment after true Ieaves form on the weeds. A postemergence application should be made before weeds reach a height of 1 or 2 inches.

PRECAUTIONS: Not for sale or use in the U.S. Perennial weeds are not controlled. Dry weather conditions reduce the herbicidal activity. If good moisture conditions are not prevalent, irrigate as soon as possible after preemergence treatment. Make no more than two applications per year. Not effective preemergence on high organic matter soils. May cause temporary burning to crops when applied postemergence, but they will outgrow this. Grasses with the exception of annual bluegrass are poorly controlled.

ADDITIONAL INFORMATION: Phytotoxic effects are manifested by growth inhibition and chloratic and necrotic effects on weeds. More effective for the control of weeds after the cotyledon have opened. Often mixed with an adjuvant to increase the postemergence activity on weeds. Does not readily leach with water. Water solubility 4 ppm. Weed control lasts for 3-4 weeks.

NAMES

FLUOMETURON, COTORAN, METURON, HIGALECOTON, COTTONER

N,N-dimethyl-N'-[3-(trifluoromethyl) phenyl] urea

TYPE: Fluometuron is a substituted urea and is used as a selective, preemergence and postemergence herbicide.

ORIGIN: 1960. ClBA-Geigy Corporation. Also produced by other formulators.

TOXICITY: LD_{50} - 6416 mg/kg. May cause slight eye irritation.

FORMULATIONS: 4L, 85 DF. Formulated in combination with MSMA.

USES: Cotton. Used on these and other crops outside the U.S.

IMPORTANT WEEDS CONTROLLED: Cocklebur, Florida pusley, lambsquarters, morningglory, pigweed, ragweed, smartweed, barnyardgrass, Brachiaria, crabgrass, crowfootgrass, goosegrass, fall panicum, foxtail, ryegrass, teaweed, purslane, sesbania, jimsonweed, and others.

RATES: Applied at 1-1.5 kg ai/ha.

APPLICATION: Used as a preemergence, early postemergence, or layby treatment. Surfactants may enhance the postemergence herbicidal activity. For best results, weeds should be less than 2 inches high. Delay postemergence sprays to cotton at least until it is 3 inches tall.

PRECAUTIONS: Do not use on cotton planted in furrows. Perennial weeds are not controlled. Sensitive crops include sugar beets, soybeans, snapbeans, cole crops, tomatoes, legumes, and cucurbits. Do not plant any crop but cotton within 6 months of the last application.

ADDITIONAL INFORMATION: Adequate soil moisture enhances the herbicidal activity. Where dry weather conditions prevail, the herbicidal activity may be delayed and reduced. Enters weeds primarily through their roots, and to a lesser extent through the foliage. Solubility in water 105 ppm. May be applied by air. Gives 6-8 weeks weed control. May be mixed with siquid nitrogen solution or other herbicides.

RELATED MIXTURES:

1. CROAK—A combination of fluometuron and MSMA developed by Drexel Chemical Co. for postemergence weed control in cotton.

RELATED COMPOUNDS:

1. *METHOPROTRYNE*, GESARAN — A substituted urea compound developed by Ciba Geigy in 1965 which is used outside the U.S. as a postemergence herbicide on cereals to control grasses.

NAMES

CHLORBROMURON, MALORAN

3-(4-bromo-3-chlorophenyl)-1 -methoxy-1-methylurea

TYPE: Chlorbromuron is a urea compound used as a selective, pre and early postemergence herbicide.

ORIGIN: 1961. ClBA-Geigy Corporation.

TOXICITY: LD_{50} - 5000 mg/kg. May cause slight eye and skin irritation.

FORMULATION: 50% WP.

166

USES: Used outside the U.S. on peas, celery, sunflowers, potatoes, soybeans and carrots.

IMPORTANT WEEDS CONTROLLED: Barnyardgrass, nightshade, cocklebur, henbit, teaweed, chickweeds, fiddleneck, velvetleaf, carpetweed, crabgrass, foxtails, goosegrass, jimsonweed, lambsquarters, pigweed, purslane, ragweed, smartweed, mustard, and others.

RATES: Applied at .75-2.5 kg a.i./ha.

APPLICATION: Applied as a preemergence treatment before weeds appear. Used postemergence on celery and carrots.

PRECAUTIONS: Not for sale or use in the U.S. Do not use on soybeans planted in deep furrows or planted Iess than 1 3/4 inches deep. Do not use on light soils. Heavy rainfall may result in crop injury. Do not apply to soils where organic matter is Iess than 1% or over 5%. Perennial weeds are not controlled.

ADDITIONAL INFORMATION: Disturb the soil as little as possible for 5 weeks after application. Solubility in water 35 ppm. May be tank mixed with other herbicides. Closely related to linuron. Sufficient moisture is required to carry this material into the rootzone of germinating weeds. Taken up by both the roots and Ieaves. Gives 6-8 weeks weed control.

NAMES

METOBROMURON, PATORAN, PATTONEX

3-(4-bromophenyl)-1-methyloxy-1-methylurea

TYPE: Metobromuron is a substituted urea compound used as a selective, preemergence herbicide.

ORIGIN: 1964. ClBA-Geigy Corporation.

TOXICITY: LD_{50} - 2000 mg/kg. May cause eye and skin irritation.

FORMULATION: 50% WP. Formulated with other herbicides.

USES: Used outside the U.S. on potatoes, beans, soybeans, peppers, sunflowers, tomatoes, and tobacco.

IMPORTANT WEEDS CONTROLLED: Lambsquarters, smartweeds, bluegrass, pigweed, purslane, chickweed, mustards, shepherd 's purse, jimsonweed groundcherry, nightshade, galinsoga, matricaria, and others.

RATES: Applied at 1.5-2.5 kg ai/ha.

APPLICATION: Applied as a preemergence herbicide immediately after planting.

PRECAUTIONS: Not for sale or use in the U.S . Some bean and tobacco varieties will be injured. Certain crops will be injured if planted within one year following application of this material . Slightly toxic to fish. Do not use on extremely sandy soils or high organic matter soils.

ADDITIONAL INFORMATION: Water solubility 330 ppm. Most effective on broadleaf species. Taken up by the plant mostly through the root system. Gives approximately 8 weeks' weed control. Moisture is required to move the material into the soil.

NAMES

CHLOROTOLURON, CLORTOKEM, DICURAN, TOLUREX, TRIMARAN, LUDORUM, TORO, TAUSMAN

N-(3,Chloro-4-methylphenyl)-N-N-dimethylurea

TYPE: Chlorotoluron is a urea compound used as a selective, pre and postemergence herbicide.

ORIGIN: 1969. CIBA-Geigy Corporation.

TOXICITY: LD_{50} - 10,000 mg/kg.

FORMULATIONS: 80% WP, 20G, 50 SC. Formulated with other herbicides.

USES: Being used outside the U.S. on cereal grains.

IMPORTANT WEEDS CONTROLLED: Wild oats, ryegrass, bluegrass, blackgrass, quackgrass, shepherd's purse, lambsquarters, smartweed, chickweed, and many others.

RATES: Applied at 1.5-3 kg ai/ha.

APPLICATION: Applied either preemergence or early postemergence to cereals while weeds are young and actively growing. Rainfall is required to carry it into the soil. If applied postemergence, use on cereals from the 3-leaf stage to tillering.

PRECAUTIONS: Not for sale or use in the U.S. Some varieties of wheat and barley are injured.

ADDITIONAL INFORMATION: Solubility in water 70 ppm. No residue remains in the soil to injure the crops planted in the fall. Very effective on both grass and broadleaves. Non-toxic to bees. May be applied by air. Taken up by both leaves and roots.

NAMES

LINURON, LOROX, AFALON, SIOLCID, LINUREX ROTALIN, SARCLEX, LINEX

(3,4-dichlorophenyl)-1-methoxy-1-methylurea

TYPE: Linuron is a substituted urea used as a selective herbicide, applied both pre and postemergence.

ORIGIN: 1960. E.l. DuPont de Nemours & Company, and Hoechst AG.

TOXICITY: LD_{50} - 1500 mg/kg. May irritate eyes, nose, throat, and skin.

FORMULATION: 50% WP, 50 DF, 50 DG, 4 F.

USES: Corn, soybeans, sorghums, asparagus, carrots, celery, parsnips, potatoes, and winter wheat. Used for total vegetation control at higher rates. Used outside the U.S. on these and sugar beets, beans, leeks, peas, grapes, ornamentals, and other crops.

IMPORTANT WEEDS CONTROLLED: Barnyardgrass, crabgrass, foxtail, mustard, lambsquarters, pigweed, purslane, ragweed, and many other annual weeds. Not effective on perennial weeds.

RATES: Applied at .5-3 Ib actual/A.

APPLICATION: Applied as a preemergence herbicide and as a directed postemergence application. Rainfall or overhead irrigation is required to take it into the soil. May be used postemergence on emerged seedling weeds that are 2-6 inches tall.

PRECAUTIONS: Crop injury may result on sandy soils or soils containing less than 1% organic matter. If extremely heavy rains follow application, crop injury may result. Do not apply over the top of corn.

ADDITIONAL INFORMATION: Aerial application may be made on some crops. Sprayed fields may be replanted if seeds fail to produce a satisfactory stand, but do not retreat. Often mixed with other herbicides. Water solubility 75 ppm. Postemergence activity on small annual weeds if applied during periods of high temperature and high humidity. May be used with liquid fertilizers.

NAMES

DIFENOXURON, LIRONION

N-4-(p-methoxyphenoxy)-phenyl-N,N-dimethylurea

TYPE: Difenoxuron is a substituted urea compound used as a selective, postemergence herbicide.

ORIGIN: 1964. CIBA-Geigy Corporation.

TOXICITY: LD_{50} - 7750 mg/kg.

FORMULATION: 50% WP.

USES: Used outside the U.S. on leeks, garlic, and onions.

IMPORTANT WEEDS CONTROLLED: Annual broadleaf weeds.

RATES: Applied at 2.5 kg active/ha.

APPLICATION: Apply as a selective herbicide on onions when they are at the crook stage or older. Best activity is when weeds are in the 4-leaf stage.

PRECAUTIONS: Not for sale or use in the U.S. Not effective on grasses.

ADDITIONAL INFORMATION: Water solubility 20 ppm . Taken up by leaves and roots. Weed control lasts for 3-4 weeks.

NAMES

MONOLINURON, ARESIN, ARRESIN

N-(4-chlorophenyl)-N-methoxy-N-methylurea

TYPE: Monolinuron is a urea compound used a a selective, pre and postemergence herbicide.

ORIGIN: 1963. Hoechst AG of Germany.

FORMULATIONS: 50% WP. 200 EC. Also formulated with other herbicides.

TOXICITY: LD_{50} - 1430 mg/kg.

USES: Outside the U.S. on potatoes, corn, grapes, beans, asparagus, berries, ornamentals and other crops.

IMPORTANT WEEDS CONTROLLED: Many annual broadleaves and grasses. Spectrum is about the same as linuron.

RATES: Applied at .5-3 kg a.i./ha.

APPLICATION: Applied both pre and postemergence to the weeds. However, postemergence applications should be made while the weeds are still young. Applied to potatoes prior to their emergence. Apply to damp soil as a preemergence application.

PRECAUTIONS: Not sold or used in the U.S. Evaporates relatively easily. Do not use on soils high in organic matter.

ADDITIONAL INFORMATION: Uptake is both through the roots and leaves. The compound has a relatively high vapor pressure as compared to other urea compounds, so uptake via the stomata is possible. Kills plants by the inhibihon of photosynthesis. May persist in the soil. Water solubility 735 ppm.

NAMES

DIMEFURON, PRADONE, RANGER, SCORPIO

2-tert. butyl-4-(2-chloro-4-(3,3-dimethylureido) phenyl)-
1,3,4-oxadiazolin-5-one

TYPE: Dimefuron is an urea compound used as a selective pre and postemergence herbicide.

ORIGIN: 1974. Rhone Poulenc.

TOXICITY: LD_{50} - 1000 mg/kg.

FORMULATIONS: Sold formulated with carbetamide as a 70% WP.

USES: Outside the U.S. on oilseed rape, cabbage, clover, peas, beans, peanuts, cotton, cereals, and alfalfa. At high rates it is used for total vegetation control.

IMPORTANT WEEDS CONTROLLED: Annual grasses and broadleaves.

RATES: Applied at .2-2 kg a.i./ha.

172

APPLICATION: Applied both pre and postemergence. Rainfall is required to move it into the soil.

PRECAUTIONS: Not for sale or use in the U.S.

ADDITIONAL INFORMATION: Very effective against broadleaf weeds. Control will last for up to 18 weeks.

NAMES

BENZTHIAZURON, GANON, GATNON

1-methyl-3-(2-benzthiazolyl)-urea

TYPE: Benzthiazuron is a selective urea compound used as a preemergence herbicide.

ORIGIN: 1966. Bayer AG of Germany.

TOXICITY: LD_{50} -1000 mg/kg. Some eye irritation may occur.

FORMULATION: 80% WP.

USES: Outside the U.S. on sugar beets, spinach, and fodder beets.

IMPORTANT WEEDS CONTROLLED: Pimpernel, corn spurry, annual bluegrass, mustard, shepherd's purse, galinsoga, wild radish, poppies, nettle, sowthistle, chickweed, lambsquarters, and many others.

RATES: Applied at 3.2-6.4 kg a.i./ha..

APPLICATION: Spray only on soils that are free of weeds, 4-5 days after seeding. Preplant applications may also be made. Rainfall soon after treatment is required to move the chemical into the weed germination zone. If applied preplant, incorporate lightly into the soil with a harrow or similar equipment. This method gives the best control in dry climates.

173

PRECAUTIONS: Not for use in the U.S . Do not use on peat or muck soils or soils high in organic mauer. Weeds not controlled include dock, knotweed, foxtail, wild oats, and others . Do not mix with PCNB or Disyston. Growing weeds will not be controlled. Do not apply once beets have emerged.

ADDITIONAL INFORMATION: Weeds are controlled by being absorbed through the root system as they germinate. Moisture is required to activate the chemical. Makes a colorless, non-staining spray solution. Remains effective for several weeks, but soil residues should disappear in time to plant the next crop. Grain, potatoes, corn, or beets can be planted should the treated beet crop be lost due to bad weather conditions. More active on broadleaf weeds than on grasses. Solubility in water 12 ppm.

NAMES

METHYLDYMRON, STACKER

I-(alpha, alpha-dimethylbenzyl)-3-methyl-3-phenyl urea

TYPE: Methyldymron is a urea compound used as a selective, preemergence herbicide.

ORIGIN: 1982. SDS BIOTECH of Japan.

TOXICITY: LD_{50} - 3948 mg/kg.

FORMULATIONS: 50% WP.

USES: Used in Japan on turf. Experimentally being tested on rice, corn, cotton, beans, sunflowers, peanuts, sugarcane, potatoes, strawberries, and others.

IMPORTANT WEEDS CONTROLLED: Cyperacious species as well as barnyardgrass and annual bluegrass.

RATES: Applied at 2-10 kg a.i./ha.

APPLICATION: Applied as a preemergence application immediately.

PRECAUTIONS: Not for sale or use in the U.S.

ADDITIONAL INFORMATION: Water solubility 120 ppm. Absorbed by the roots and translocated in the plant.

NAMES

DAIMURON, DYMRON, SHOWRONE

1-(alpha, alpha-dimethylbenzyl) -3-p-totyl urea

TYPE: Dymron is a urea compound used as a selective, preemergence herbicide.

ORIGIN: 1981. SDS Biotech of Japan.

TOXICITY: LD_{50} - 4000 mg/kg.

FORMULATIONS: 80% WP, 7% granules. Also formulated with other herbicides.

USES: Rice in Japan.

IMPORTANT WEEDS CONTROLLED: Cyperaceous weeds and grasses.

RATES: Applied at .45-2 kg a.i./ha.

APPLICATION: Apply to paddy rice fields at the time of or shortly after sowing or transplanting of the crop.

PRECAUTIONS: Not for sale or use in the U.S.

ADDITIONAL INFORMATION: Water solubility 1.7 ppm. Preemergence activity only. Rice is extremely tolerant to this product. Absorbed by the roots and rapidly translocated.

NAMES

BROMACIL, HYVAR-X, URAGAN, ROKAR, HYVAR-L

CH₃—C⟨NH⟩C=O
Br—C—C(=O)—N—CH—CH₂—CH₃
                    |
                   CH₃

5-bromo-6-methyl-3-(1-methoxypropyl)-2,4 (1H, 3H) pyrimidinedione

TYPE: Bromacil is a substituted uracil compound used as a pre and postemergence herbicide.

ORIGIN: 1963. E.I. DuPont de Nemours & Company.

TOXICITY: LD_{50} - 5200 mg/kg. May cause skin irritation.

FORMULATIONS: 3.2 lb/gal. 80% WP, 2 Ib/gal water-miscible concentrates, 10% pellets.

USES: Non-crop areas, pineapples, and citrus.

IMPORTANT WEEDS CONTROLLED: Foxtail, crabgrass, cheatgrass, ragweed, brush species, lambsquarters, pigweed, quackgrass, dandelion, plantain, johnsongrass, bermudagrass, dallisgrass, nutgrass, bracken fem, and many other perennial and annual weeds.

RATES: Applied at 2-30 lb a.i./acre.

APPLICATION: Apply just before or during the weeds' active growing period. Rainfall is necessary to carry this material into the root zone so plants can take it up. Remove dense vegetation for the best results. Agitate while spraying. Apply to citrus at least 4 years old. May be applied to pineapple immediately after planting.

PRECAUTIONS: Unsatisfactory results have been obtained when applied late in the season after plants have hardened off. In areas where less than four inches of rainfall occurs during the growing season, unsatisfactory control of deeprooted perennial weeds can be expected. Do not apply when the ground is frozen. Do not apply near desired plants. Avoid drift. Do not plant any but tolerant crops for 2 years after treatment.

ADDITIONAL INFORMATION: A quicker kill results, if applied during the early part of the growing season. Particularly effective on perennial grasses. Slow acting. Non-corrosive and non-volatile. Use higher rates on soils high in carbon or organic matter. Absorbed by the roots. Seasonal or longer control can be expected. Must be carried into the root zone by moisture. Lower rates may be used when applied as a preemergence treatment. The use of a surfactant greatly increases its contact activity. A small amount of rainfall (1/2 inch) is enough to activate this material. Water solubility 815 ppm. Relatively non-toxic to birds and fish.

RELATED MIXTURES:

1. KROVAR I—A combination of 40% bromacil and 40% diuron developed by DuPont Co. for use on citrus and for total vegetation control.

2. KROVAR II—A combination of 53% bromacil and 25% diuron developed by DuPont Co. for use on citrus and for total vegetation control.

NAMES

TERBACIL, GEONTER, SINBAR

5-chloro-3-(1,1-dimethylethyl)-6-methyl-2,4-(1H, 3H)-pyrimidinedione

TYPE: Terbacil is a uracil compound used as a selective, preemergence herbicide.

ORIGIN: 1962. E.l. DuPont de Nemours and Company.

TOXICITY: LD_{50}- 1082 mg/kg. May cause slight eye irritation.

FORMULATIONS: 80% WP.

USES: Sugarcane, apples, peaches, alfalfa, blueberries, asparagus, caneberries, mint, and pecans. Used on these and other crops outside the U.S.

IMPORTANT WEEDS CONTROLLED: Most annual grasses and broadleaves. Higher rates are generally needed for large-seeded annuals such as wild oats. Shallow-rooted perennials such as quackgrass may be controlled with higher rates.

RATES: Applied at .4-8 lbs a.i./acre.

APPLICATION: Apply as a preemergence treatment or during the very early stages of seedling development. Use the lower rates on light, sandy soils in areas of low seasonal rainfall. Rainfall or sprinkler irrigation is required following application to move this material into the weed germination zone.

PRECAUTIONS: Do not use on sandy soils or soils containing less than 1% organic matter. Do not plant non-tolerant crops in treated area for 2 years. Apply only to trees that have been established for 2-3 years.

ADDITIONAL INFORMATION: Control can be expected for a period of 2-4 months.

NAMES

LENACIL, ADOL, VENZAR, VISOR, VIZOR

3-cyclohexyl-5,6-trimethyleneuracil

TYPE: Lenacil is a uracil compound used a a preplant or preemergence, selective herbicide.

ORIGIN: 1964. E.l.-DuPont de Nemours and Company.

TOXICITY: LD_{50} - 11,000 mg/kg. May cause eye irritation.

FORMULATION: 80% WP. Formulated with other herbicides.

USES: Used outside the U.S. on sugar beets, fodder beets, flax, spinach, cereals, strawberries, and ornamentals.

IMPORTANT WEEDS CONTROLLED: Lambsquarters, smartweed, wild radish, knotweed, wild buckwheat, crowfoot, pennycress, chickweed, shepherd's purse, foxtails, pineapple weed, groundsel, corn spurry, and others.

RATES: Applied at .4-2 kg a.i./ha.

APPLICATION: Applied as a preplant incorporated, preemergence or postemergence application.

PRECAUTIONS: Not used in the U.S. Use lower rates on light soils. Injury may result on sandy or gravel soils with low organic matter contents. Soil incorporate on peat or muck soils to obtain good weed control. Speedwells, wild oats, bedstraw, pigweed, and all perennial weeds will not be controlled. Exceptionally heavy rains following application may result in crop injury.

ADDITIONAL INFORMATION: Moisture is required to move the material into the weed germination zone. Unhealthy crop growth from environmental factors may decrease the beets' tolerance to this compound. Beet injury symptoms are stunting or vein clearing of lhe young seedlings. Weed control can be expected for a 2 month period. Emerged weeds will not be controlled.

NAMES

METHABENZTHIAZURON, TRIBUNIL

N-2-benzothiazolyl-N,N-dimethylurea

TYPE: Methabenzthiazuron is a urea compound used as a selective, postemergence and preemergence herbicide.

ORIGIN: 1967. Bayer AG of Germany.

TOXICITY: LD_{50} - 1000 mg/kg.

FORMULATION: 70% WP. Formulated with other herbicides.

USES: Outside the U.S. on winter and spring sown cereals, peas, garlic, potatoes, orchards, vineyards, beans, onions, and artichokes.

IMPORTANT WEEDS CONTROLLED: Annual meadowgrass, corngrass, blackgrass, silky bentgrass, pigweed, shepherd's purse, goosefoot, barnyardgrass, fumitory, nettle, henbit, knotweed, wild radish, nightshade, sowthistle, chickweed, speedwells, and many others.

RATES: Applied at 1.4-2.8 kg a.i./ha.

APPLICATION: On winter-sown cereals, apply in spring as soon as machines can be driven over the fields. On spring-sown cereals, apply when they are in the 4-leaf stage. Weeds should be small and growing actively at the time of application. Also applied preemergence within 8 days after planting.

PRECAUTIONS: Do not apply in combination with urea or other liquid fertilizers. Not used in the U.S. Do not use on cereals under-sown with clover. Avoid drift. Cereals showing signs of winter kill should not be sprayed. Do not use on light, sandy soils or those high in organic matter.

ADDITIONAL INFORMATION: Soil should be moist at the time of application. Primarily taken up through the roots. It takes 14-20 days for the treated weeds to die. Root-propagated weeds are not controlled. Does not last in the soil long enough to effect the following crop, since the residual activity is 3-6 months. Synergistic with phenoxy-type compounds.

NAMES

METOXURON, DEFTOR, DOSAFLO, DOSANEX, PIRUVEL, PURIVEL, SULEREX

N-(3-chloro-4-methoxyphenyl)-n-n-dimethyl urea

TYPE: Metoxuron is a urea compound being used as a postemergence, selective herbicide.

180

ORIGIN: 1968. Sandoz Ltd. of Switzerland.

TOXICITY: LD_{50}- 2000 mg/kg.

FORMULATIONS: 80% WP, 80% WG. Formulated with other herbicides.

USES: Outside the U.S. on carrots and cereals (winter wheat, winter rye, winter barley). Also used for preharvest defoliation on potatoes, flax, and hemp.

IMPORTANT WEEDS CONTROLLED: Grasses (blackgrass, silky bentgrass, wild oats, ryegrass) and most annual broadleaves (mayweeds, etc.)

RATES: Applied at 2.4-4 kg a.i./ha.

APPLICATION: On cereals, apply postemergence when grain is in the 3-leaf stage up to the end of tillering. A spring treatment is suggested. On carrots, apply when carrots have 3 true leaves, or preemergence.

PRECAUTIONS: Do not use on any food or feed crop in the U.S. Injury results on some cereal varieties.

ADDITIONAL INFORMATION: A short-residual herbicide. Non-toxic to bees and fish. Water solubility is 23 ppm.

NAMES

ISOPROTURON, ARELON, GRAMINON, SABRE, TOLKAN, ALON, PASPORT, ISOTOP, SWING, HYTANE, AUGUR, FALIREXON, PROTUGAN

N-(4-isopropylphenyl)-N,N-dimethylurea

TYPE: Isoproturon is a urea derivate used as a selective herbicide in pre and postemergence applications.

ORIGIN: 1974. Hoechst AG of Germany. CIBA-Geigy, Shell Agar, Rhone-Poulenc and others market the product.

TOXICITY: LD_{50} - 1826 mg/kg.

FORMULATIONS: 50% WP, 4 F. Formulated with other herbicides.

IMPORTANT WEEDS CONTROLLED: Blackgrass, wild oats, ryegrass, annual meadowgrass, and many annual broadleaved weeds.

USES: Outside the U.S. on cereals.

RATES: 1-2 kg/ha ai.

APPLICATION: Applied preemergence in the fall or postemergence in the spring when grain is in the 3-leaf stage until the end of tillering. Activity is increased by a high soil moisture content.

PRECAUTIONS: Do not incorporate into the soil. Not being developed in the U.S.

ADDITIONAL INFORMATION: Solubility in water: 70 ppm. Compatible with other herbicides. Absorbs through lhe root system as well as the leaves. May be applied by air.

NAMES

TEBUTHIURON, **BRULAN, TIUROLAN, HERBEC, PREFLAN, TEBULAN, SPIKE, GRASLAN, BUSHWACKER, COMBINE, SCRUBMASTER, HERBIC**

N-(5-(1,1-dimethylethyl)-1,3,4-thiadiazol)-2-yl)-N,N-dimethylurea

TYPE: Tebuthiuron is a substituted, urea-type compound used as a pre and postemergence herbicide.

ORIGIN: 1972. Elanco Products Co. Now marketed by DowElanco.

TOXICITY: LD_{50}- 644 mg/kg.

FORMULATIONS: 80% WP, 5% granules, 10%, 20%,40% pellets.

IMPORTANT WEEDS CONTROLLED: Alfalfa, bluegrasses, bromes, bouncingbet, buttercups, chickweed, clovers, cocklebur, dock, fescue, fiddleneck, filaree, foxtails, goldenrod, henbit, horseweed, kochia, lambsquarters, morningglory, mullein, nightshade, wild oats, pigweed, puncturevine, ryegrass, prickly sida, sowthistle, spurge, sunflower, Russian thistle, vetch, witchgrass, woody plants, and many others. Many brush species, trees, and perennial broadleaf plants are also controlled.

USES: Total vegetation control and rangeland and pastureland. Sold outside the U.S. on sugarcane.

RATES: Applied at .75-10 lb a.i./ha.

APPLICATION:

1. Total Vegetation Control—Apply as either a pre or postemergence treatment to railroad right-of-ways, industrial sites, tank farms, highway medians, etc. Rainfall is required to move the chemical into Ihe soil. Apply either before or during the period of active plant growth. May be applied under asphalt.

2. Sugarcane—Applied to both plant and stubble cane. Apply both pre and postemergence.

3. Rangeland—The pellets are used to control brush species while the grasses continue to grow. Used in long-range brush control programs. Control is complete in 12-24 months.

PRECAUTIONS: Do not apply near desirable trees or plants, or where their roots will grow into the treated areas. Do not apply to any portion of a ditch bank that will come in direct contact with water.

ADDITIONAL INFORMATION: May be used in combination with other herbicides. Controls a number of woody species of plants. Use higher rates and/or repeat application to control hard-to-kill perennial species. Spot treatment on range and pasture land is effective. Water solubilily 2.5 ppm. Low toxicity to fish and wildlife. Vertical leaching in the soil is slow and no lateral chemical movement has been observed. Readily absorbed through the root system and inhibits photosynthesis. The half life in the soil is 12-15 months in areas of 40-60 inches of rainfall. Woody plants take a period of 2-3 years to be completely controlled.

ISOURON, ISOXYL

3-(5-tert-butyl-3-isoxazoyl)-1,1-dimethyl urea

TYPE: Isouron is a urea compound used as a selective, pre and postemergence herbicide.

ORIGIN: 1980. Shionogi Co. of Japan. Being marketed by Shell Chemical Intl. in certain areas of the world.

TOXICITY: LD_{50} - 630 mg/kg.

FORMULATIONS: 50% WP, 4% granules.

USES: Used outside the U.S. on sugarcane and as a non-crop herbicide.

IMPORTANT WEEDS CONTROLLED: Most annual broadleaf weeds and grasses as well as some perennials.

RATES: Applied at 2.5-10 kg/ha. Use the lower rates on sugarcane.

APPLICATION: Apply either pre or postemergence to the weeds. Rainfall must take the product into the root zone.

PRECAUTIONS: Not for sale or use in the U.S. Do not apply near desired plants.

ADDITIONAL INFORMATION: Herbicidal effects are relatively slow in appearance. Absorbed through the roots. Water solubility .7 ppm. Non-corrosive.

METAL ORGANICS AND INORGANICS

NAMES

DSMA, ANSAR-8100, DSMA LIQUID, METHAR

$$CH_3-AS \begin{array}{c} O \\ \| \\ \end{array} \begin{array}{c} O-Na \\ O-Na \end{array}$$

Disodium methanearsonate

TYPE: DSMA is an organic arsenical compound used a a selective, postemergence, contact herbicide.

ORIGIN: 1956. The Ansul Co., Pamol, ISK Biotech, and Drexel are among the principle basic producers today.

TOXICITY: LD_{50} - 600 mg/kg. May cause some skin irritation.

FORMULATIONS: 3.6 SC, 81 SP, 7.2 lb/gl F.

IMPORTANT WEEDS CONTROLLED: Crabgrass, johnsongrass, dallisgrass, nutgrass, foxtail, cocklebur, barnyardgrass, witchgrass, velvetgrass, chickweed, goosegrass, knoxweed, and others.

USES: Cotton, citrus, non-crop areas, and turf.

RATES: Applied at 2-4 Ib actual/A.

APPLICATION:

Cotton—Apply as a directed spray, after the cotton is 3-4 inches high, up to first bloom. A second treatment 1-2 weeks after the first may be necessary. Do not make more than 2 applications per season. A slight burning or reddening may result on the cotton plant, but it quickly outgrows this. On turf, apply to young crabgrass when it is 1-2 inches high (2-3 -leaf stage). 2-4 applications spaced about a week apart may be necessary to control the crabgrass as it emerges. Use a higher rate on larger crabgrass and dallisgrass. Keep the turf grass roots moist after application. Do not treat new turf until after 3 mowings.

PRECAUTIONS: Do not apply to cotton on after the first bloom. Do not apply to St. Augustine grass, bahiagrass, or centipedegrass turf. Fescues and bentgrass may be injured if appied during hot weather. Avoid drift. All plants are injured by high concentrations.

ADDITIONAL INFORMATION: Non-corrosive. A very soluble compound, so it leaches through the soil readily. Reduce rates during periods of high temperatures (above 85°F) to avoid turf injury. No preemergence activity. Compatible with 2,4-D. Most effective on grasses. Cotton appears to be slightly more tolerant to this material than to MSMA. Most effective on young, rapidly growing weeds when temperatures are above 70°F.

NAMES

MSMA, ANSAR 529, BUENO, DACONATE, MERGE, MESAMATE, WEED-HOE-108, TARGET, ARSONATE, PUEDEMAS

$$CH_3 - \overset{\displaystyle O}{\overset{\displaystyle \|}{\underset{\displaystyle \underset{\displaystyle Na}{|}}{AS}}} - OH$$

Monosodium acid methanearsonate

TYPE: MSMA is an organic arsenical compound used as a selective contact herbicide applied postemergence.

ORIGIN: 1956. The Ansul Co. ISK Biotech, Pamol, Drexel, and Vineland Chemical Company are among the principle basic producers today.

TOXICITY: LD_{50} - 700 mg/kg. Slightly irritating to the skin and eyes.

FORMULATIONS: 4-8 lb/gal solutions.

USES: Cotton, citrus, agricultural plantings, non-crop areas and turf. Used outside the U.S. on plantation crops, sugarcane, citrus and fruit trees.

IMPORTANT WEEDS CONTROLLED: Johnsongrass, nutgrass, crabgrass, barnyardgrass, goosegrass, dallisgrass, cocklebur, and many others.

RATES: Apply at 2-5 lb actual/A.

APPLICATION:

Cotton—Apply as a directed spray only when cotton is above 3-4 inches high until the first bloom. A second treatment 1-3 weeks after the first application may be necessary. A slight burning or red discoloration may occur to cotton, but it will readily outgrow the effects. May be applied preplant if cotton planting was delayed and weeds have emerged.

188

On turf applied as a postemergence application to control crabgrass and other weeds. Two applications are usually required at 7-10 day intervals. Do not use on St. Augustine, bahiagrass or centipede grass turf. May be used as a directed spray on citrus and nonbearing fruit trees.

PRECAUTIONS: When used on cotton, keep foliage contact at a minimum. Do not apply to newly planted turf until after the 3 mowings.

ADDITIONAL INFORMATION: Closely resembles DSMA in its action, only it is slightly more phytotoxic. More effective than DSMA under high temperatures. Topical sprays do not result in excessive phytotoxicity to cotton when applied prior to first bloom. Temperature should be above 70°F to be the most effective.

RELATED MIXTURES:

1. BROADSIDE, MONCIDE—A combination of 3 lb MSMA and 1.25 lb cacodylic acid per gallon marketed by Drexel Chemical Co. and Monterey Chemical Co. for weed control in noncrop areas.

2. QUADMEC, TRIMEC PLUS—A combination of MSMA, 2,4-D, MCPP and dicamba developed by PBI Gordon to be used on turf.

NAMES

CACODYLIC ACID, BOLLS-EYE, CLEAN-BOLL, COTTON-AIDE, ERASE, MONTAR, PHYTAR-560, LEAF-ALL, QUICK-PICK

Dimethylarsenic acid (sodium salt) plus sodium cacodylate

TYPE: Cacodylic acid is an organic arsenical compound used as a non-selective, postemergence, contact herbicide.

ORIGIN: 1958. Developed by the Ansul Co. Produced today by Pamol Ltd. Formulated by Drexel Chemical, Monterey Chemical and Platte Chemical Co.

TOXICITY: LD_{50} - 830 mg/kg.

189

FORMULATIONS: 2.48-3.25 lb/gal solution.

USES: Non-crop areas, cotton, and ornamental plants.

IMPORTANT WEEDS CONTROLLED: Nutgrass, dallisgrass, crabgrass, johnsongrass, bermudagrass, spurge, pigweed, purslane, lambsquarters, morningglory, Russian thistle, puncturevine, dodder, and many others.

RATES: Applied at 2.5 to 7.5 lb actual.

APPLICATION: For general weed control, add a surfactant to get maximum spreading of the material. Spray foliage thoroughly. The younger the weed growth, the better the kill will be. Use a directed spray around trees and shrubs. For best results, spray when the temperature exceeds 70°F. On cotton used as a preharvest defoliant. Apply 7-10 days prior to picking, when 50% of the bolls are open.

PRECAUTIONS: Avoid drift. Mildly corrosive.

ADDITIONAL INFORMATION: Control is obtained only through foliar absorption. Results are apparent within 5 days and retreatment will be necessary if there is green growth after this period. Non-staining and non-corrosive. Leaves no harmful residue in the soil. Effectively used to kill grass under trees since it is not root-absorbed. Woody plants are killed by frill injection. The higher the temperatures, the greater the activity. May be mixed with other defoliants when used on cotton.

NAMES

SODIUM CHLORATE, CHLORAX, DEFOL, DE-FOL-ATE, DROP-LEAF, FALL, HARVEST AID, KLOREX, KNOX-EM-OFF, KUSATOL, LEAFEX, TUMBLEAF, DERVAN

Na Cl O3

TYPE: Sodium chlorate is an inorganic salt that non-selectively destroys germinating seeds and inhibits plant growth.

ORIGIN: 1910. Occidental, Drexel, and Kerr McGee are among the major basic producers.

TOXICITY: LD_{50} - 1200 mg/kg. Somewhat irritating to the skin, eyes, and mucous membranes.

FORMULATIONS: Numerous formulations are available, 99% active material in powder form. 3-6 Ib/gal flowable and 2, 3, 4 Ib/gal aqueous solution. Formulated with other herbicides.

USES: Mainly used on non-crop land for spot treatment and for total vegetation control on roadsides, fenceways, ditches, etc. As a defoliant and desiccant for use on cotton, safflower, corn, flax, peppers, soybeans, grain sorghum, southern peas, dry beans, rice and sunflowers.

IMPORTANT WEEDS CONTROLLED: Johnsongrass, bermudagrass, paragrass, quackgrass, bindweed, Canada thistle, leafy spurge, Russian knapweed, and most other annual and perennial weeds.

RATES: Applied at 1-3 Ib/100 sq ft for total vegetation control. For desiccation apply at 6-9 lb/acre.

APPLICATION: Most effective when applied directly to the soil as dry crystals, or in a spray solution for total vegetation control. As a desiccant apply 7-14 days prior to harvest.

PRECAUTIONS: A strong oxidizing agent, therefore highly inflammable. Clothing and vegetation contaminated with chlorate or its solution are dangerously flammable. Corrosive to zinc and steel, especially in the presence of moisture. Dangerously reactive with compounds capable of oxidation such as organic substances, solvents, oils, sulfur, sulfides, phosphorous, ammonium salts,and others. Do not use around buildings because of the fire hazard. This chemical has a salty taste and salt-hungry animals may eat enough to become poisoned. Do not use near desired plants. Avoid drift.

ADDITIONAL INFORMATION: Persists in the soil. Toxicity in soil is decreased considerably by a high nitrate content, alkaline conditions, and high soil temperatures. Non-volatile at ordinary temperature. Non-odorous. Plants absorb through both roots and leaves. It is carried downward through the xylem since it kills the phloem tissue. It increases the rate of respiration, decreasing calalase activity, depleting the plant's food reserves. Chlorate injured plants are more susceptible to frost. Takes up moisture when in contact with damp air. Borates are added to many formulations since they reduce the fire hazard 30-50 times more toxic to plants than sodium chloride (table salt). Decomposition occurs more readily in moist soils above 70°F. Dormant seeds in the soil usually survive the treatment.

RELATED COMPOUNDS:
AMMONIUM SULFAMATE, AMMATE—An older compound developed by DuPont for brush control in non-crop areas. It is no longer used in the U.S. but still used for this purpose in other parts of the world.

NAMES

BORAX, BORASCU, PYROPOR, TRONABOR, POLYBOR, COMAC

$Na_2 B_4 O_7$ 10 H2O
Various mixtures are available

TYPE: Borax is a non-selective, inorganic salt used for total vegetation control .

ORIGIN: First used around 1920-1930. U.S. Borax. Formulated today by a number of different companies.

TOXICITY: LD_{50} - 2500 mg/kg. May cause skin irritation.

FORMULATIONS: 100%granules and crystals. Mixed with other herbicides or sodium chlorate.

USES: Widely used for long-term total vegetation control.

IMPORTANT WEEDS CONTROLLED: Bindweed, Canada thislle, leafy spurge, horsenettle, whiletop, Russian knapweed, wild rose, sumac, trumpe vine, honeysuckle, bur ragweed, johnsongrass, poison oak and ivy, bermudagrass, paragrass, quackgrass, klamalhweed, and most other annual and perennial plants.

RATES: Applied at 400-6000 lb actual/A.

APPLICATION: Apply evenly to the surface of the area to be treated. Requires about 1 inch of rain to wash it into the root zone. This chemical can be applied at any time of the year; however, it works best on annual weeds when they are young and rapidly growing. It becomes highly stable in the soil, working on the deepest weed root systems. Lower rates can be used where annual weeds are the only problem. Perennial grasses should be cut off close to the ground before treatment for the best results. May be used under asphalt.

PRECAUTIONS: Do not get near desired plants. Renders the entire area unproductive for at least a year.

ADDITIONAL INFORMATION: Can be mixed with asphalt to keep weeds from growing through it. Decomposes very slowly in the soil. Causes the plants to yellow and dry up (desiccate) . Usually effective for a number of years. Most often mixed with other herbicides. Phytotoxicity is in direct proportion to the boron content. Slow acting. Persists for longer periods in heavy clays and/or alkaline soils. However, the toxicity may be reduced due to alkaline soil reaction which fixes boron in relatively non-available forms. A non-flammable and non-corrosive material. Plants take it in through their root systems.

RELATED MIXTURES:

1. UREABOR—A non-selective soil herbicide, containing 66.5% sodium tetraborate, 1.5% bromacil and 30% sodium chlorate, sold by Simplot Chemical Co.

2. MONOBOR CHLORATE or MBC—A non-selective soil herbicide, containing 68% sodium metaborate and 30% sodium chlorate, sold by J. R. Simplot Chemical Co.

3. BARACIDE—A non-selective soil herbicide sold by Security Chemical Co. containing 66.5% sodium metaborate, 30% sodium chlorate and 1.5% bromacil.

4. LESCOBOR—A non-selective soil herbicide sold by Lesco, Inc. containing 66.5% sodium metaborate, 30% sodium chlorate and 1.5% bromacil.

5. BOROCIL IV—A non-selective herbicide sold by J.R. Simplot Co., containing 94% sodium metaborate and 4% bromacil.

6. BARESPOT WEED AND GRASS KILLER—A non-seleclive herbicide sold by J.R. Simplot Co., containing 66.5% sodium metaborate, 30% sodium chlorate and 1.25% diuron.

NAMES

COPPER SULFATE, **BLUESTONE**

$CuSO_4 5H_2O$

Copper sulfate pentahydrate

TYPE: Copper sulfate is an inorganic copper compound used as an algicide.

ORIGIN: First used around 1930. Griffin, Agtrol, Phelps Dodge and others are the principle formulators today.

TOXICITY: LD_{50}- 300 mg/kg.

FORMULATIONS: Copper sulfate is 25.5% copper and formulated in different size crystals.

USES: Used for algae control in ponds, potable water, fish hatcheries, fish ponds, rice fields, lakes, streams, ditches, and water supplies. Also used as a wood preservative, in anti-fouling paints, and as a fungicide.

IMPORTANT WEEDS CONTROLLED: Most species of algae.

RATES: Usually applied at .1-11 ppm. With some algae species when used at .15 ppm, repeated daily for 3-5 days, it gives better control than a single large dosage. To rice, apply at up 15 lbs/A.

APPLICATION: Apply over the water surface above the weeds in the early summer. May be repeated in the late summer. May be used in a cloth bag and put where water flows over it or tow behind a boat until it dissolves. Also may be dumped into flowing water which will take it downstream. To rice apply uniformly over the fields by air after flooding for algae control.

PRECAUTIONS: High rates are toxic to fish. Burns the foliage of plants. Corrosive to metals. Incompatible with soap. Do not apply to water below 60°F. Do not use in water containing trout if the carbonate hardness of the water does not exceed 50 ppm.

ADDITIONAL INFORMATION: Treated areas will turn a grayish-white soon after treatment. The water will be free of algae within several days. Maintain a concentration of 1 ppm early in the growing season reducing gradually through the summer to .6 ppm. Fish tolerate 1 ppm, and it is safe to use as irrigation water at that concentration. Must be applied in sunlight to have effective results. Control should last five weeks or longer. Swimming and fishing are allowed immediately after treatment. Often mixed with Diquat for aquatic weed control. Early treatment gives the best control.

RELATED COMPOUNDS AND MIXTURES:

1. KOMEEN—A copper ethylenediamine compound produced by Griffin Corp., used to control aquatic weeds, such as hydrilla and algae.

2. K-TEA—A copper-triethanolamine complex produced by Griffin Corp. used to control hydrilla and algae in water systems.

3. CUTRINE-PLUS—A 9% copper derived from copper-ethanolamine complex marketed by Applied Biochemists to control algae and hydrilla in crop and non-crop irrigation water, potable water, reservoirs, farm and fish ponds, and in lakes, and fish hatcheries. A 3.7% granules is also available.

4. *ZINC CHLORIDE*, MOSS-KILL—An inorganic compound used in the homeowner market to control moss and algae on walls, roofs, patios and decks. Sold by a number of formulators.

5. *FERRIC SULFATE, FERROUS SULFATE,* MOSS-OUT—An inorganic compound sold by a number of companies into the homeowner market to control moss and algae in lawn and turf areas.

6. AGRIBROM—A biocide developed by Great Lakes Chemical to control algae by contact in greenhouses. Sold as a powder, granule or tablet.

7. ALGAE-RHAP—A copper triethanolamine complex compound used as an aquatic herbicide to control algae that was developed by Agtrol Chemical Products.

OTHER HERBICIDES

Petroleum Derivatives, Phosphates, Carbothiolates, Cyclic Compounds, Halogenated Hydrocarbons, Aliphatic Acids, and others.

NAMES

PETROLEUM OILS, CONTACT OILS, DIESEL OIL, FORTIFIED OILS, FUEL OIL, KEROSENE, PETROLEUM SOLVENTS, STOVE OIL, STODDARD SOLVENT, WEED OILS

TYPE: Weed oils are both selective and non-selective, contact herbicides applied postemergence or preemergence before the crop emerges.

ORIGIN: Made by many manufacturers, mostly oil companies. Being replaced by newer compounds.

TOXICITY: Considered non-toxic. May cause some skin and eye irritation.

FORMULATIONS: 100% petroleum compounds or various dilutions.

USES: Non crop weed control.

IMPORTANT WEEDS CONTROLLED: Most weeds upon contact.

APPLICATION: Usually applied preemergence or as a postemergence spray.

PRECAUTIONS: Do not apply on desired plants unless otherwise recommended. Avoid drift. Mostly being replaced by safer, easier to use compounds.

ADDITIONAL INFORMATION: Viscosity influences the rate at which an oil will spread over and penetrate a plant. The higher the viscosity, the slower the penetration. Most contact oils have a flash point of over 180°F. Most selective oils have a flash point of over 100°F. Oils for weed control should contain a high percentage of unsaturated hydrocarbon and have a sulfonatable residue of at least 25%. Those used selectively are usually more highly refined, so that unpleasant odors or taste in market crops are reduced or eliminated. Oils also express insecticidal activity.

NAMES

FATTY ACIDS, POTASSIUM SALTS, SHARP SHOOTER, DEMOSS, ERASER

Formula not given

TYPE: The potassium salt of fatty acids are organic compounds used as a non selective contact herbicide.

ORIGIN: 1988 Safer Inc. Now being sold and developed by Mycogen and Ringer Corp.

TOXICITY: 18% and 40% a.i. liquid.

USES: Used on walkways, driveways, flowerbeds, trees and shrubs to control weeds. Used on roofs, decks, walkways, greenhouses, trees, turf, etc. to control mosses.

IMPORTANT WEEDS CONTROLLED: Mosses, algae, lichens, liverworts, annual and perennial weeds.

RATES: For moss control apply 4-6 fl. oz/gallon of water. For weed control, apply 22-44 oz/gallon of water.

APPLICATION: Apply to the point of runoff to all unwanted vegetation. Most effective on young succulent weeds less than 5" high. On turf areas for moss control water the moss and surrounding grass prior to treatment. Apply thoroughly and do not water for 2 days.

PRECAUTION: May foam in the sprayer. Do not apply when rainfall is expected. Treated surfaces may be temporarily slippery. Do not use on turf when temperatures exceed 85°F. Avoid drift.

ADDITIONAL INFORMATION: Mosses and algae after application will dry and slough off naturally. Works faster in warmer temperature. Rapid acting.

NAME

MCDS, ENQUIK

Monocarbamide dihydrogen sulfate

TYPE: MCDS is a monourea sulfuric acid compound used as a contact herbicide.

ORIGIN: 1986. Unocal Corp. (Div. of Union Oil)

TOXICITY: LD_{50} - 350 mg/kg. Causes severe eye and skin irritation

FORMULATION: 81.6% liquid.

USES: Used as a herbicide on onions, leeks, shallots, garlic, vegetable row middles, onions, peanuts, and grass seed. Used as a desiccant on potato vines.

IMPORTANT WEEDS CONTROLLED: Most annual weeds, especially broadleaves will be controlled.

RATES: Applied at 15-20 gallons/A in 20-50 gallons of water.

APPLICATION: Applied postemergence to crops and weeds as a contact herbicide. Directed or shielded sprays may be used on susceptible crops.

PRECAUTIONS: Caustic and corrosive. Explosively decomposes above 160°F. Do not mix with nitrogen fertilizers. Corrosive to nylon, aluminum and any copper alloy such as brass Mount spray booms on the rear of the tractor to reduce corrosion. Do not treat if rain is expected within 12 hours. Avoid drift. Do not apply by air. Do not use with a surfactant if selectivity is desired. Do not use with flood type nozzle.

ADDITIONAL INFORMATION: Destroys plant tissue by disrupting the cell membrane structure. Fast acting. No systemic activity. Each gallon, alter decomposition in the soil, produces up to 1.9 pounds of available nitrogen and 2 lbs of available sulfur. Selectivity is obtained because of the waxy cuticle on the leaf of some plants which prevents wetting. No residual activity.

NAMES

ACROLEIN, MAGNACIDE-H, AQUALIN

$CH_2=CH-OH$

2-propenal acrylaldehyde

TYPE: Acrolein is a hydrocarbon used as a contact aquatic herbicide.

ORIGIN: Shell Chemical Co., 1956. Now manufactured for Magna Chemical Co.

TOXICITY: LD_{50} 10.3 mg/kg. Extremely toxic to eyes and skin.

FORMULATION: 92% liquid (6.5 lb a.i./gal)

USES: Used as an aquatic herbicide in moving water.

IMPORTANT WEEDS CONTROLLED: Most submersed and floating aquatic weeds and algae.

RATES: Applied at 1-15 ppm or at .166-1.5 gallons per CFS (cubic feet/second). The rate is dependent upon amount and size of the aquatic weeds and the water temperature. The more weeds the more product is needed. If water temperatures are between 60°-55° F increase rate by 20%, if 55°-50° F increase by 50% and if below 50° increase by 100%.

APPLICATION: Injection into the water over a period of 15 minutes to 8 hours, to form a wave of treated water controlling the weeds as it moves. The faster flowing the canal, the longer the application period. Apply as soon as weed growth appears, and repeat at 2-3 weeks intervals if regrowth occurs.

PRECAUTIONS: Do not exceed the rate of 15ppm. Flammable. Corrosive. Fish, shrimp and crabs will be killed at the application rates. Do not release any water for 6 days after application into fish rearing waters.

ADDITIONAL INFORMATION: Emerged aquatic weeds such as tules and cattails are not controlled. Weeds die and desintegrate slowly over a period of 3-4 days to 2 weeks. Floating aquatic weeds are controlled but it generally takes the higher rates. For the best control, the water temperature should be above 60°F.

NAMES

DEVINE, P.p.

Phytophthora palmivora MWV disease.

TYPE: DeVine is a plant disease being used as a biological herbicide.

ORIGIN: 1976. Abbott Labs.

TOXICITY: Non-toxic.

IMPORTANT WEEDS CONTROLLED: Strangler or milkweed vine.

USES: Being used in citrus in Florida, since this vine will get in the groves and completely kill the citrus trees.

RATES: Applied at 1/2 pt of product/A.

APPLICATION: Apply only to a soil in the area where the vine is a problem. Up to 90% control can be obtained within 1-2 years of a single application. The soil must be wet at the time of application.

PRECAUTIONS: Do not apply to dry soil . Do not tank mix with other pesticides; however, they can be applied prior to or immediately following an application. The formulation must be handled as a perishable commodity. Must be used within 4 weeks of it manufacture. Do not apply to areas where susceptible plants will later be grown. Do not apply with chlorinated city water. Keep the product refrigerated.

ADDITIONAL INFORMATION: This is the first microbial herbicide to be put on the market. Control is obtained in 2-10 weeks. Produced by submerged fermentation and must be stored under refrigeration (2-8°C). Dying milkweed vine plants are girdled at the soil line and up to an inch above it and the plant dies from the root infection. Does not survive in the soil except in association with infected milkweed vine root debris. Citrus trees are not susceptible to this disease, but it will infect watermelon and periwinkle.

NAME

COLLEGO

Living spores of the fungus Collectotrichum gloesporioidies spp. aeschynomene

TYPE: Collego is a biological fungus being used as a herbicide.

ORIGIN: 1981. The Upjohn Co. Being marketed and developed by Ecogen Inc.

TOXICITY: LD_{50} - 5000 mg/kg. May cause eye irritation.

FORMULATIONS: 15% active product (75.7 x 10^{10} viable fungal spray).

IMPORTANT WEEDS CONTROLLED: Northern jointvetch, (curly indigo).

USES: Rice and soybeans.

APPLICATION: Apply when northern jointvetch is 8-24 inches tall but has not reached the bloom stage. The leaves of the weed should be wet and expected to remain so for 12 hours. Relative humidity should be above 80% and the temperature above 80°F for 12 hours after application. This is a 2-component product. Component A consists of a water-soluble rehydrating agent that allows the spores to take up water prior to germination. Component B is the dried fungal spores. Usually applied by air.

PRECAUTIONS: Do not apply when rice or soybeans are under stress for moisture. Do not apply to crops previously treated with phenoxy herbicides. Do not apply fungicides for 3 weeks following application. Chemical pesticide residues left in the spray tank may kill the live Collego spores so clean thoroughly before using with activated charcoal. Do not allow the product to remain in the spray tank for over 12 hours or allow it to heat up. Incompatible with fertilizer, insecticides, fungicides, and herbicides. Store at temperatures of 40°F to 80°F.

ADDITIONAL INFORMATION: The product will cause disease lesions that will completely encircle the stems of the northern jointvetch plant. Diseased plants will become limp and may collapse. If they are not killed, they will become weak and will not be able to produce seed. Death to the plants may take 5 weeks after application. Compatible with Blazer herbicide. If disease lesions do not encircle the stems within 14 days of application, a second treatment may be necessary.

NAME

BIO MAL

Colletotrichum gloeosporioides sp. malvae

TYPE: Bio Mal is a naturally occuring fungi that is used as a bioherbicide.

ORIGIN: Philom Bros. of Canada. 1989. Being marketed in Canada by DowElanco.

TOXICITY: Non toxic to warm blooded animals. May cause eye and skin irritation.

FORMULATIONS: WP

IMPORTANT WEED CONTROLLED: Round-leaved mallow.

USES: Registered for use in Canada on barley, buckwheat, canola, flax, lentils, mustard, oats, rye, sugarbeets, sunflower, soybeans and wheat.

APPLICATION: Applied with common application equipment. Apply after the mallow is in the two leaf stage, preferably before it is over 15 cm. tall. Effective if applied later but control is slower. For best results, apply in high humidity. (80-100%). Rainfall during or after application will increase the efficacy.

PRECAUTIONS: If humid conditions do not occur poor control may be observed and a second application may be necessary. Do not apply by air. Use within 3 hours of mixing with water. Do not mix with other pesticides. Agitation is necessary. Sealed packages should be stored frozen.

ADDITIONAL INFORMATION: Based on a naturally occuring fungi that attacks the mallow plant. Lesions form on the stem of the plant 2-4 weeks after application that develops into open wounds. Complete control takes 3-6 weeks. Second growth is also controlled. Once the mallow has been destroyed, the fungal spores decline to their natural level. The formulation is the dried spores of the fungus. Gives season long control from one application.

NAME

CASST

Alternaria cassiae

TYPE: CASST is a bioherbicide that is used for the selective control of sicklepod.

ORIGIN: 1986. Mycogen.

TOXICITY: Non-toxic to warm blooded animals.

FORMULATION: 100% active ingredient.

USES: Experimentally being tested on soybeans and peanuts.

IMPORTANT WEEDS CONTROLLED: Sicklepod, and coffee senna.

APPLICATION: Applied as a postemergence treatment.

PRECAUTIONS: Used on an experimental basis only.

ADDITIONAL INFORMATION: Efficiency is influenced by the amount of moisture present on the plant surfaces. Surfactants may enhance the effectiveness. Combination with other herbicides and growth regulators are under investigation.

NAME

MYX -1200

Fusarium lateritium

TYPE: MYX-1200 is a bioherbicide that is used for the selective control of velvetleaf.

ORIGIN: 1987. Mycogen.

TOXICITY: Non-toxic to warm blooded animals.

FORMULATION: Under investigation.

USES: Collon and soybeans, on an experimental basis.

IMPORTANT WEEDS CONTROLLED: Velvetleaf.

APPLICATION: Applied with chemical herbicides to enhance velvetleaf control .

PRECAUTIONS: Used on an experimental basis only.

RELATED COMPOUNDS:

I . MYX- 1621—A bioherbicide based on Colletotrichum truncatum being developed by Mycogen for the control of Florida beggarweed in peanuts and other crops.

NAMES

FLAMPROP-M-METHYL, FLAMPROP-M-ISOPROPYL, LANCER, MATAVEN, COMMANDO, GUNNER, FLAME, SUFFIX-BW, SUPER BARNON, BARNON PLUS, EFFIX

N-benzoyl-N-(3 chloro-4-fluorophenyl)DL-alanine

TYPE: Flamprop-m-methyl is an arylalanine compound used as a selective, postemergence herbicide for the control of wild oats in wheat.

ORIGIN: 1971. Shell Research Ltd. of England.

TOXICITY: LD_{50},1210 mg/kg. May cause eye and skin irritation.

FORMULATION: 105 g/l EC.

USES: Used on wheat in Australia, Canada, Mexico and Europe.

IMPORTANT WEEDS CONTROLLED: Wild oats and blackgrass.

RATES: Applied at .45-.675 kg ai/ha.

APPLICATIONS: Apply postemergence to vigorously growing wheat at the proper growth stages, from the end of tillering to the first-node stage of the crop.

PRECAUTIONS: The correct timing of application is essential. The wheat crop should be growing well and not subjected to stress. Not for sale or use in the U.S. Do not tank mix with phenoxy compounds and delay 10 days before or after their application.

ADDITIONAL INFORMATION: The material should be used only on wheat, not other small grains. Good coverage of the wild oat plant is required. May be tank mixed with fungicides.

NAMES

FLUAZIFOP-P-BUTYL, FUSILADE, FUSILADE 2000, GRASS-B-GON, ONECIDE, ORNAMEC

Butyl 2-[4-(5-trifluoromethyl-2-pyridinyl) oxy] phenoxy] propionate.

TYPE: Fluazipop-p-butyl is a selective systemic herbicide for the control of grass weeds in broadleaved crops.

ORIGIN: 1980. Ishihara Sangyo Kaisha Ltd. of Japan. ICI Plant Protection is licensed to develop the product.

TOXICITY: LD_{50} - 1490 mg/kg. May cause eye irritation.

FORMULATIONS: 1 EC.

USES: Asparagus, cotton, endive, rhubarb, spinach, sweet potatoes, stone fruits, garlic, coffee, pecans, peppers, onions, soybeans, non-bearing tree and vine crops, non-crop areas and ornamentals in the U.S. Sugar beet, oil seed rape, potatoes, cotton, soybean, groundnuts, vines, citrus fruits, coffee, onions, bulbs, bananas, peas, beans, brassicae, other vegetable, bush fruit, and plantation crops outside the U.S.

IMPORTANT WEEDS CONTROLLED: Annual and seedling perennial grasses at low rates, mature perennial grasses at higher rates, Bermudagrass, johnsongrass, quackgrass, proso millet, red rice, shattercane, volunteer corn, and others.

RATES: Applied at .125-2 kg a.i./ha.

APPLICATION: Most active when applied postemergence when weeds are in the 2-4-leaf stage. Two applications are sometimes required. Apply to johnsongrass when 12-18 inches tall. Used with a crop oil concentrate or a non ionic surfactant.

PRECAUTIONS: Broadleaf weeds and perennial sedges are tolerant. Toxic to fish. Avoid drift. Do not use if rainfall is expected within 1 hour.

ADDITIONAL INFORMATION: Fluazifop remains active in the soil and may exert an effect on susceptible crops sown up to 4 months after application of high rates (2 kg/ha). Biological activity persists longer in light, sandy soils. Translocation from the leaves to the roots. All broadleaf crops are tolerant to this material. Slow acting. May be used with other herbicides.

RELATED MIXTURES:
FUSION—A combination of fluazifop-p-butyl and fenoxaprop-ethyl developed by ICI for postemergence grass control in soybeans.

NAMES

FENOXAPROP-P-ETHYL, FENOXAPROP-ETHYL, ACCLAIM, DEPON, EXCEL, FURORE, PODIUM, LASER, CHEETAH, HORIZON, ISOMERO, OPTION, PUMA, WHIP, WHIP SUPER, BUGLE, RALON, PUMA-S

Ethyl:(±)-ethyl-2-[4-(6-chloro-2-benzoxazolyl) oxy]phenoxy]propanoate

TYPE: Fenoxaprop-ethyl is a postemergenee, selective herbicide to control annual and perennial grasses in broadleaf crops.

ORIGIN: 1982. Hoeehst AG of West Germany. Being marketed on soybeans by FMC.

TOXICITY: LD_{50} - 2357 mg/kg. Can cause slight eye and skin irritation.

FORMULATIONS: 1 EC, .79 EC. 2 EC.

USES: Rice, cotton, peanuts, soybeans, turf and non-crop areas. Being used outside the

U.S. on beans, peanuts, alfalfa, vegetables, clover, rape, turf, potatoes, sugar beets, sunflowers, tobacco, broadleaf crops, and others to control grasses.

IMPORTANT WEEDS CONTROLLED: Annual grasses, perennial grasses including johnsongrass, wild oats, crabgrass, barnyardgrass, panicum, volunteer corn, and others.

RATES: Applied at 40-480 g a.i./ha.

APPLICATION: Applied when most of the grasses have emerged. Applied to johnsongrass when it is 10-15 inches tall. A second application may be necessary under some circumstances. Annual grasses should be 4-6 inches tall at application. Rice is tolerant from the 4 leaf stage to early tillering.

PRECAUTIONS: Used on Kentucky Bluegrass grown only East of the Rocky Mountains. Do not mow treated areas for 24 hours. Rainfall within 1 hour may reduce control. No activity against Lolium, Poa or Bromus species. Quackgrass and bermudagrass are not controlled. Do not apply with Blazer or Tackle or propanil herbicides or apply either of these within 7 days of an application. Do not apply with phenoxy compounds. Toxic to fish.

ADDITIONAL INFORMATION: No soil activity. A contact herbicide that is partly systemic. Acts primarily through the foliage. Growth of the weed stops with the treatment, but symptoms may not be seen for 4-10 days. Higher rates are required in semi-arid regions. No effect on broadleaf crops has been noted. Can be used to take certain grasses out of turf species. May be applied by air. Use on only cool season turfgrass species. Use with a crop oil concentrate. Rainfall will not effect the results 1-3 hours after treatment. May be used with other herbicides.

RELATED MIXTURES:

1. TILLER—A combination of fenoxaprop, 2,4-D and MCPA being marketed by Hoechst-Roussel to be used on wheat.

2. CHEYENNE — A combination of fenoxaprop-ethyl, MCPA and a sulfonyl urea developed by Hoechst Roussel for use on wheat.

3. DAKOTA — A combination of fenoxaprop-ethyl and MCPA developed by Hoechst Roussel for use on wheat.

NAMES

QUIZALOFOP-P-ETHYL, ASSURE II, PILOT, TARGA SUPER, ZERO, SCHERIFF

Ethyl(R)-2-[4-[(6-chloro-2-quinoxalin-2-yl)oxy]-phenoxy]propionate

TYPE: Quizalofop-p-ethyl is an organic compound used as a selective postemergence herbicide.

ORIGIN: 1979. Nissan Chemical Ind. of Japan. Licensed to be developed and marketed in other countries by FBC Ltd. of England, a subsidiary of Schering AG of West Germany and Rhone Poulenc. Marketed in the U.S . by E.I. DuPont.

TOXICITY: LD_{50} -1182 mg/kg. May cause eye and skin irritation.

FORMULATIONS: .8 Ib ai EC.

USES: Soybeans and non crop areas. Being developed for use on rape, peanuts, potatoes, sugar beets, and other broadleaf crops. Used on these and other crops outside the U.S.

IMPORTANT WEEDS CONTROLLED: Bromes, couchgrass, crabgrass, volunteer cereals, foxtails, blackgrass, ryegrass, wild oats, bermudagrass, barnyardgrass, Dallisgrass, canarygrass, johnsongrass, and other grasses.

RATES: Applied at .125- 1.25 kg ai/ha. Higher rates are used on perennial grasses.

APPLICATION: Apply when the grass weeds are in the 2-10 leaf stage. Apply with a crop oil concentrate or non-ionic surfactant. The crop can be treated at any time the weeds are in the appropriate growth stage. A second application may be necessary on perennial grasses. To johnsongrass, apply before it becomes 2 feet tall. A second application may be necessary. To bermudagrass, apply before the runners are over 6 inches in Iength and on quackgrass, apply when the plants are 6-10 inches in height. In irrigated areas, apply immediately after an irrigation.

PRECAUTIONS: Does not control sedges and broadleaf weeds. Somewhat toxic to fish. Do not store below 32°F. Do not use with vegetable oil spray adjuvants. Stressed weeds will not be controlled. Avoid drift. Minor leaf spotting to soybeans may occur. Do not apply in a tank-mix with postemergence broadleaf herbicides.

ADDITIONAL INFORMATION: Quickly absorbed and translocated in the plant. Controls the root systems of perennial grasses. Rainfall 1-3 hours after application will not reduce the effectiveness. Will kill grass in 10-21 days. May be mixed with other herbicides. May be applied by air.

NAMES

QUIZALOFOP-P-TEFURYL, UBI-C4874, PANTERA

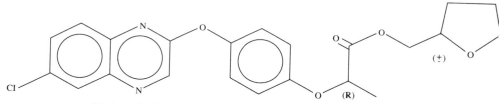

(±)-tetrahydrofurfuryl (R)-2-[4-(6-chloroquinoxalin-2-yloxy) phenoxy] propanoate

TYPE: Quizalofop-p-tefuryl is a propanoate compound used as a selective, postemergence herbicide.

ORIGIN: Uniroyal 1990.

TOXICITY: LD_{50} 1140 mg/kg. May cause slight eye irritation.

FORMULATION: 1EC, .3EC

USES: Experimentally being tested on rape, cotton, beans, flax, peas, peanuts, potatoes, soybeans, sugarbeets, sunflowers and other broadleaf crops.

IMPORTANT WEEDS CONTROLLED: Blackgrass, wild oats, bromes, sandbur, crabgrass, barnyardgrass, volunteer cereals, foxtails, panicum, quackgrass, bermudagrass, johnsongrass and others.

RATES: Applied at 30-150 g a.i./ha.

APPLICATION: Applied postemergence to the grassy weeds when they are growing actively, usually 2-4 weeks after planting the crop. On johnsongrass apply when it is at least 8 inches tall. A second application may be needed. Bermudagrass should be at least 3 inches tall and quackgrass at least 6 inches tall.

PRECAUTIONS: Used on an experimental basis only. Broadleaf weeds and sedges are not controlled.

ADDITIONAL INFORMATION: Should be used with a crop oil concentrate. Slow acting with no activity noted for 6-10 days, and complete control taking up to 21 days.

NAMES

PROPAQUIZAFOP, AGIL, SHOGUN, FALCON, CORRECT

2-isopropylidineamino-oxyethyl (R)-2-(4-(6-chloroquinoxalil-2-yloxy)phenoxy) propionate

TYPE: Propaquizafop is an organic compound used as a selective postemergence herbicide.

ORIGIN: 1985. Dr. R. Maag of Switzerland. Being developed by Ciba Geigy.

TOXICITY: LD_{50}, - 500 mg/kg. May cause eye and skin irritation.

FORMULATION: 240 EC, 1 EC.

USES: Outside the U.S. on alfalfa, cotton, sunflowers, tomatoes, vegetables, grapes, fruit trees, coffee, beans, peas, peanuts, potatoes, rape, soybeans, sugar beets and various vegetable crops.

IMPORTANT WEEDS CONTROLLED: Wild oats, bromes, crabgrass, barnyardgrass, ryegrass, panicums, foxtails, johnsongrass, quackgrass, bermudagrass and others.

RATES: Applied at 60-250 g ai/ha.

APPLICATION: Apply postemergence when grasses are young and actively growing. A split application may be necessary to control perennial weeds.

PRECAUTIONS: Not for sale or use in the U.S. Toxic to fish. Do nof use on cucurbit crops. Broadleaf weeds are not controlled.

ADDITIONAL INFORMATION: Systemic in activily. Treated grasses cease growth in 1-2 days and completely die in 10-20 days. Sedges and broadleaf weeds are not controlled. Rainfall 1 hour after treatment will not decrease the effectiveness. It may be tank-mixed with other herbicides. Most active under warm humid condition. May be applied by air.

NAMES

FLUPROPANATE, TETRAPION, FRENOCK, ORGA

Sodium-2,2,3,3-tetra fluoropropionate

TYPE: Flupropanate is an organic compound used as a selective, translocated herbicide.

ORIGIN: 1969. Daikin Kogyo Co. of Japan. Now marketed by ICI.

TOXICITY: LD_{50} - 8500 mg/kg.

FORMULATIONS: 70 and 90% SC, 6% granules.

USES: Used outside Ihe U.S. on rubber, tea, jute, cotton, sugarcane, fruit trees, pastures and non-cultivated areas.

IMPORTANT WEEDS CONTROLLED: Annual bluegrass, crabgrass, foxtails, wild oats, barnyardgrass, annual sedges, bermudagrass, couchgrass, pampasgrass, nutsedge, bulrush, and others.

RATES: Applied at 1.5-16.5 kg actual/ha.

APPLICATIONS: Applied either during the dormant or the growing season.

PRECAUTIONS: Not for sale or usage in the U.S.

ADDITIONAL INFORMATION: More effective against grasses than broadleaves. Long residual control. May be mixed with fast acting, postemergence herbicides. Relatively non-toxic to fish. Little contact activity.

NAMES

CLODINAFOP, CGA-184927, TOPIK

2-propynyl-(*R*)-2-[4-(5-chloro-3-fluoro-2-pyridyloxy)-
phenoxy] propionate

TYPE: Clodinafop is a pyridine compound used as a selective postemergence herbicide.

ORIGIN: Ciba Geigy 1990.

TOXICITY: LD_{50} 1829 mg/kg.

FORMULATION: Formulated with a safener CGA-184927 in a 4:1 ratio.

IMPORTANT WEEDS CONTROLLED: Blackgrass, wild oats, annual ryegrass, canarygrass, foxtails and others.

USES: Being developed outside the U.S. on wheat and rye.

RATES: Applied at 40-80 g a.i./ha.

APPLICATION: Applied as a postemergence herbicide when weeds are in the 3 leaf stage up to early shooting. May be used with a crop oil concentrate.

PRECAUTION: Not for sale or use in the U.S. Poa annua and perennial grasses are not controlled. Cereals do not have adequate tolerance to this product when used alone, so it must be applied with the safener. Do not use on barley.

ADDITIONAL INFORMATION: Effective under a wide variety of environmental conditions. Application timing is very flexible with this material. Performs best when grasses are actively growing and have plenty of moisture. Can be applied to wheat at any stage of growth. Susceptible weeds stop growth 1-2 days after application. Slow acting.

NAMES

CINMETHYLIN, ARGOLD, CINCH

exo-l-methyl-4-(l-methylethyl)-2-(2-methylphenylmethoxy)-7-oxabicyclo-[2,2,1] heptane

TYPE: Cinmethylin is a cineole compound used as a selective, preplant or preemergence herbicide.

ORIGIN: 1982. Shell Chemical Co.

TOXICITY: LD_{50}, - 3960 mg/kg. May cause eye and skin irritation.

FORMULATION: 1 EC, 1.5 and 3% granules.

USES: Being used outside the U.S. on rice.

IMPORTANT WEEDS CONTROLLED: Barnyardgrass, foxtails, crabgrass, panicum, millets, and others.

RATES: Applied at 75-100 g ai/ha.

APPLICATION: Applied either preplant incorporated or as a preemergence treatment. To rice, apply 3-14 days after transplanting with barnyardgrass is in the 1-2.5 leaf stage. On directed seeded rice, apply 6-8 days after sowing.

PRECAUTIONS: Not for use in the U.S. Somewhat toxic to fish.

ADDITIONAL INFORMATION: Water solubilily 61 ppm. More effective against grasses than broadleaves. Some activity against nutsedge. Inhibits meristematic growth in both roots and shoots. Gives 4-8 weeks control. May be mixed with other pesticides. Most effective under high temperatures.

NAMES

CHLORNITROFEN, MO, CNP

2,4,6-trichlorophenyl-4-nitrophenyl ether

TYPE: Chlornitrofen is a diphenyl ester used as a selective herbicide.

ORIGIN: 1965. Mitsui Toatsu Chemical Industry Co. Ltd. of Japan.

TOXICITY: LD_{50}- 10,000 mg/kg.

FORMULATIONS: 9 and 18% granules, 20% EC.

IMPORTANT WEEDS CONTROLLED: Numerous weed species.

USES: In Japan on direct sowed paddy rice, upland rice, transplanted rice, carrots, cabbage, and lettuce.

APPLICATION: Apply from 3-6 days after transplanting. On other crops use as a preemergence treatment.

PRECAUTIONS: Do not use for other than the recommended uses. Store in a dry, cool, area. May cause injury when applied to wet foliage. Not for sale or use in the U.S.

ADDITIONAL INFORMATION: A contact herbicide. Closely related to nitrofen. Non-toxic to fish. Does not suppress germination, but kills weeds after germination. Doesn't suppress root development. Control is effective for about 30 days.

NAMES

ACIFLUORFEN-SODIUM, TACKLE, BLAZER

Sodium 5-(2-chloro-4-(trifluoromethyl)-phenoxy)-2-nitrobenzoate

TYPE: Acifluorfen-sodium is a diphenyl ether compound used as a selective, preemergence and postemergence herbicide.

ORIGIN: 1975. Rohm & Haas Company and Mobil Chemical Co. Now being marketed by BASF and Rhone Poulenc.

TOXICITY: LD_{50}, - 1370 mg/kg. Irritating to the eyes and skin.

FORMULATION: 2EC, 2L.

USES: Soybeans, peanuts, and rice. Being sold outside the U.S. on a number of crops.

IMPORTANT WEEDS CONTROLLED: Cocklebur, ragweed, coffeeweed, morningglory, mustard, pigweed, lambsquarters, smartweed, velvetleaf, jimsonweed, and others.

RATES: Applied at .25-1 lb active/A.

APPLICATION: Applied as an over-the-top postemergence application. Apply when weeds are in the 2-4-leaf stage. For preemergence control, higher rates are required. Soybeans are tolerant at the cotyledon stage up to mature beans, but application should not be made from the 1-3-trifoliate leaf stage. May be applied by air.

PRECAUTIONS: Do not use in periods of dry weather. Do not soil incorporate. Rain or irrigation within 6 hours of application will reduce the effectiveness. Do not apply when the temperature is below 60°F. Does not control grasses.

ADDITIONAL INFORMATION: Most effective postemergence. Soybeans show good tolerance although some effects may be seen on new leaf growth which are readily outgrown. Applied to the crop at any growth stage. Burns back the top growth of perennial weeds. Fast acting since the weeds will be dead in 3-5 days. More active in bright sunlight. May be applied with a nitrogen solution to increase velvetleaf control in soybeans. May be mixed with other herbicides.

RELATED MIXTURES:

I . GALAXY—A combination of acifluorfen-sodium and bentazon to be used postemergence on soybeans, marketed by BASF.

2. STORM—A combination of acifluorfen-sodium and bentazon developed by BASF as a postemergence herbicide on soybeans.

NAMES

OXYFLUORFEN, GOAL, KOLTAR

2-chloro-1 -(3-ethoxy-4-nitrophenoxy)-4-trifluoromethyl benzene

TYPE: Oxyfluorfen is a diphenylether compound used as a selective preemergence and postemergence herbicide.

ORIGIN: 1974. Rohm & Haas Company.

TOXICITY: LD_{50}- 5000 mg/kg.

FORMULATIONS: 1.6EC.

IMPORTANT WEEDS CONTROLLED: Lambsquarters, purslane, ragweed, ground-sel, henbit, malva, jimsonweed, smartweed, prickly sida, pigweed, velvetleaf, mustard, barnyardgrass, crabgrass, foxtails, seedling johnsongrass, and many others.

USES: Artichokes, broccoli, cabbage, cauliflower, citrus, coffee, conifers, cotton, ornamentals, guava, jojoba, mint, onions, horseradish, tree, nut and vine crops. Used outside the U.S. on these and others. Used to control witchweed in corn.

RATES: Applied at .25-2 Ib actual/A.

APPLICATION: Applied after planting, but prior to crop emergence. Also used as a postemergence spray when weeds are about 3 inches tall. To trees and vines, apply while dormant. Do not disturb the soil surface once the application has been made. Applied postemergence to onions after 2 true leaves have developed.

PRECAUTIONS: Toxic to fish. Do not use on muck or peat soils. Do not graze treated area. Do not apply to grapes less than 3 years old. Avoid drift.

ADDITIONAL INFORMATION: May be used in combination with a number of other herbicides. Stronger on broadleaves than grasses. Most active postemergence when the weeds are small. Water solubility .1 ppm. Kills weeds as they come into contact with the material during emergence. Controls malva up to10 inches high.

NAMES

LACTOFEN, COBRA

I-(carboethoxy) ethyl 5-[2-chloro-4-(trifluoromethyl)
phenoxy]-2-nitrobenzoate

TYPE: Lactofen is a diphenyl compound used as a selective postemergence herbicide with some preemergence activity.

ORIGIN: 1980. PPG Industries. Being marketed by Valent Chemical Co.

TOXICITY: LD_{50} - 5960 mg/kg. May cause eye and skin irritation.

FORMULATIONS: 2EC.

USES: Soybeans, conifers and cotton. Experimentally being tested on beans, potatoes, peanuts, rice, and other crops.

IMPORTANT WEEDS CONTROLLED: Pigweed, nightshade, mustard, ragweed, jimsonweed, purslane, groundcherry, teaweed, sunflower, coffeeweed, velvetleaf, carpetweed, cocklebur, copperleafs, beggarweed, spurge, and others. Perennial weeds are not controlled.

RATES: Applied at .1-2 kg. ai/ha.

APPLICATION: Apply one (1) application on soybeans when the weeds are in the 2-6-leaf stage about 2-3 weeks after planting. Soybean application should be made prior to the 3rd trifoliate leaf stage. On cotton, apply as a postdirected spray when the plant is at least 8 inches tall. Apply with an adjuvant or crop oil concentration.

PRECAUTIONS: Do not apply under periods of stress when the weeds are not actively growing. Do not use low pressure coarse spray delivery systems. Grasses are not controlled. Do not apply to dusty plants, as weed control will be reduced. Used on cotton only as a directed spray.

ADDITIONAL INFORMATION: Residual activity is observed on sensitive weeds from a postemergence application. May be mixed with other pesticides. Some soybean injury will be noted at time of application, but they will quickly outgrow it. Apply to weeds that are small and actively growing. Weeds only partially controlled include smartweed, sicklepod, lambsquarters, and spurred anoda. Appears to have a half-life in the soil of 1-2 months. Rainfall within 30 minutes will not reduce the effectiveness. May be applied by air.

NAME

BENZOFENAP, YUKAWIDE, MY-71

2-[4-(2,4-dicloro-m-toluoyl)-1,3-dimethyl-pyrazol-5-yloxy]4-methylacetophenone

TYPE: Benzofenap is a pyrazole compound used either as a selective preemergence or an early postemergence herbicide.

ORIGIN: 1981. Mitsubishi Petrochemical Co., Ltd. of Japan.

TOXICITY: LD_{50} - 15,000 mg/ kg.

FORMULATION: 8% granules. Formulated with other herbicides.

USES: Paddy rice.

IMPORTANT WEEDS CONTROLLED: Bulrush, waterplaintain, arrowhead, and other annual and perennial broadleaf weeds.

RATES: Applied at 1.2-2.4 kg ai/ha.

APPLICATION: Applied immediately to 3 days after transplanting of rice.

PRECAUTIONS: Not for sale or use in the U.S. Not effective against barnyardgrass or sedges.

ADDITIONAL INFORMATION: Excellent selective herbicide for rice. Slow acting. This product is not temperature dependent. Mostly used with other herbicides.

HALOXYFOP-METHYL, FOCUS, ZELLEK, GALLANT, VERDICT, ELOGE

Methyl 2-(4-((3-chloro-5-trifluoromethyl)2-pyridinyl)oxy)phenoxy)propanoate

TYPE: Haloxyfop-methyl is a pyridinyl compound used as a selective, postemergence herbicide.

ORIGIN: 1981. Dow Chemical Co. Now marketed by DowElanco.

TOXICITY: LD_{50} - 2179 mg/kg. Irritating to the eyes and may cause some skin irritation

FORMULATION: 2EC.

IMPORTANT WEEDS CONTROLLED: Quackgrass, wild oats, volunteer grains, bermudagrass, crabgrass, barnyardgrass, panicum, millets, foxtails, johnsongrass, and others. All annual grasses and most perennial grasses are suspected to be sensitive.

USES: Used outside the U.S. on alfalfa, cotton, peas, potatoes, cucurbits, beans, peanuts, rape, soybeans, sugar beets, tobacco, tomatoes, and others.

RATES: Applied at .125-.5 Ib ai/A. Must be used with a crop oil, crop oil concentrate, or a non-ionic surfactant.

APPLICATION: To annual grasses it may be applied up to the 8 inch tall stage. On johnsongrass, apply when plants are 12 inches tall, up to the boot stage. On quackgrass, apply when plants are 4-8 inches tall. On bermudagrass, apply when plants are 6 inches in height or stolon length.

PRECAUTIONS: Not for sale or use in the U.S. Sedges and broadleaf weeds are not controlled.

ADDITIONAL INFORMATION: Selectivity is expected on all broadleaf crops. Residual activity is sufficient for the control of late germinating grasses. Preemergence activity can be achieved but higher rates are required. May be tank-mixed with postemergence broadleaf herbicides.

NAMES

MOLINATE, ORDRAM

$$CH_3 - CH_2 - S - \overset{\overset{O}{\|}}{C} - N \bigcirc$$

S-ethylhexahydro-lH-azepine-1-carbothioate

TYPE: Molinate is a thiocarbanate compound used as a selective, preemergence herbicide.

ORIGIN: 1964. Stauffer Chemical Company. ICI is the basic producer today.

TOXICITY: LD_{50} - 369 mg/kg.

FORMULATION: 8EC, 10% granules. Sometimes formulated in combination with fertilizers and other herbicides.

USES: Rice.

IMPORTANT WEEDS CONTROLLED: Barnyardgrass, dayflower. Suppresses annual sedges and spikerush.

RATES: Applied at 2-3 Ibs actual/A.

APPLICATION: Used as a preplant incorporated treatment and as a postemergence treatment at preflood, during flooding or postflood. May be applied by air.

PRECAUTIONS: Once applied, a continuous water cover must be maintained. Do not contaminate domestic waters. A very odorous material.

ADDITIONAL INFORMATION: Moisture is required to activate this material. Season-long control may be expected. Rice is extremely tolerant to this material . Drift hazard is not a problem. A light discing or spike-tooth harrowing will effectively incorporate this material. The sooner the material is incorporated into the soil, the better will be the control. May be applied in combination with fertilizer. May be used on water seeded or dry-seeded rice.

RELATED COMPOUNDS:
1. ARROSOLO — A combination of molinate and propanil developed by ICI for usage on rice.

NAMES

DIMEPIPERATE, MY-93, YUKAMATE

S-1-methyl-1-phenylethyl piperidine-1-carbothioate

TYPE: Dimepiperate is a thiocarbamate compound used as a preemergence and early postemergence selective herbicide.

ORIGIN: 1979. Mitsubishi Petrochemical Co. of Japan.

TOXICITY: LD_{50} - 946 mg/kg.

FORMULATIONS: 7% granules, 50% EC.

USES: Rice outside the U.S.

IMPORTANT WEEDS CONTROLLED: Barnyardgrass.

RATES: 30-40 kg.ha (7% granules).

APPLICATION: Applied either preemergence or postemergence before the barnyardgrass reaches the 2 leaf stage. It can be used in all types of rice culture.

PRECAUTIONS: Not for sale or use in the U.S.

ADDITIONAL INFORMATION: Control will last for approximately 20 days. Since it only controls barnyardgrass it is being formulated with other herbicides to increase its spectrum. Relatively non-toxic to fish.

DITHIOPYR, DIMENSION, STAKEOUT, LAZO, LYTON, KALCORN, WE-HOPE

S,S-dimethyl 2-(difluoromethyl)-4-(2-methylpropyl)-6-trifluoromethyl)
-3,5-pyridinedicarbothioate

TYPE: Dithiopyr is a pyridine compound used as a selective pre and postemergence herbicide.

ORIGIN: 1987. Monsanto.

TOXICITY: LD_{50} - 5000 mg/kg. May cause slight eye and skin irritation.

FORMULATION: I EC, .25-1% granules. Formulated with other herbicides.

USES: Turf. Experimentally being tested on rice, peanuts, soybeans, ornamentals, sunflowers, sugarcane, cotton, trees and nut crops, citrus and others. Used in Japan on rice.

IMPORTANT WEEDS CONTROLLED: Barnyardgrass, foxtails, crabgrass, goosegrass, johnsongrass, millets, shatter cane, bluegrass, cheatgrass, lambsquarters, downy brome, mustard, teaweed, spurge, wild oats and others.

RATES: Applied at .25-1.5 lb ai/acre.

APPLICATION: To turf applied preemergence or early postemergence to the germinating weeds. On rice apply as a granule to the water after transplanting.

PRECAUTIONS: Tolerant weeds include nightshade, bindweed, Canadian thistle, kochia, pigweed and others. Store above 40°F. Do not reseed or sprig within 3 months of application. Use on turf that has been established. Toxic to fish. Avoid drift.

ADDITIONAL INFORMATION: More active on grasses than on broadleaves. Has postemergence activity. Can be used on both warm and cool season turf. May be tank mixed with other herbicides and pesticides. Water solubility is .7 ppm. Does not prevent weed seed germination. Absorbed by both the roots and the shoots of the weeds. Gives 2-3 months control. May be tank mixed with fertilizers.

NAMES

THIAZOPYR, MON-13200, DICTRAN

Methyl 2-difluoromethyl-4-isobutyl-5-(4,5-dihydro-2-thiazolyl)-6-trifluromethyl-3-pyridinecarboxylate

TYPE: Thiazopyr is a pyridine compound used as a selective preemergence herbicide.

ORIGIN: Monsanto 1990.

TOXICITY: LD_{50} 5000 mg/kg. May cause slight eye irritation.

FORMULATION: 2EC, 50% WDG, 5%G.

IMPORTANT WEEDS CONTROLLED: Annual bluegrass, barnyardgrass, bromes, crabgrass, foxtails, ryegrass, johnsongrass, wild oats, panicums, sprangletop, shattercane, sowthistle, nightshade, chuckweed, curly dock, spurge, henbit, shepardspurse, mustards and others.

USES: Experimentally being tested on cotton, sunflower, sugarcane, corn, peanuts, alfalfa, potatoes, soybeans, grapes, tree crops and others.

RATES: Applied at .1-.56 kg a.i./ha.

APPLICATION: Applied as a preplant incorporated treatment or as a preemergence treatment.

PRECAUTION: Used on an experimental basis only. Moderately toxic to fish. Persistant in the soil, so crop rotation is currently under investigation.

ADDITIONAL INFORMATION: More effective against grasses than broadleaves. Water solubility is 2.5 ppm. Seed germination is not inhibited but cell division does not proceed normally. May be combined with other herbicides.

NAMES

FLUMETSULAM, DE-498, BROADSTRIKE

N-[2,6-difluorophenyl]-5-methyl-(1,2,4) triazolo-
[1,5a]-pyrididine-2-sulfonamide

TYPE: Flumetsulam is a pyrimidine compound used as a selective preplant herbicide.

ORIGIN: DowElanco 1990.

TOXICITY: LD_{50} 5000 mg/kg. May cause slight eye irritation.

FORMULATION: Formulated in combination with trifluralin. Containing 3.65 lb a.i./gal. (3.4 lb trifluralin + .25 lb flumetsulan).

USES: Experimentally being tested on soybeans.

IMPORTANT WEEDS CONTROLLED: Beggarweed, jimsonweed, lambsquarters, mustard, nightshade, pigweed, teaweed, smartweed, sunflower, velvetleaf and most annual grasses.

RATES: Applied at 2 pint formulation/acre (2.27 l/ha).

APPLICATION: Applied as a preplant soil incorporated treatment.

PRECAUTION: Used on an experimental basis only.

ADDITIONAL INFORMATION: Rates should be increased if the soil organic matter is over 5%. Incorporate within 24 hours of application. Herbicidal activity increases as the soil pH increases.

NAMES

BROMOFENOXIM, FANERON

3,5-dibromo-4-hydroxybenzaldehyde,(2',4'-dinitrophenyl) oxime

TYPE: Bromofenoxim is an diphenyl compound used as a selective, postemergence, contact herbicide.

ORIGIN: 1969. ClBA-Geigy Corporation.

TOXICITY: LD_{50}- 1217 mg/kg.

FORMULATIONS: 50% SC. Formulated with other herbicides.

USES: Used on small grains outside of the U.S.

IMPORTANT WEEDS CONTROLLED: Mustard, shepherd's purse, lambsquarters, smartweed, nightshade, chickweed, and many others.

RATES: Applied at 1-2.5 kg ai/ha.

APPLICATION: Apply to cereals at any time after emergence. Weeds should be between emergence and the 4-5 leaf stage.

PRECAUTIONS: Not used in the U.S. Grasses and sedges are not controlled. Toxic to fish. Spray volume should not exceed 400 kg/ha.

228

ADDITIONAL INFORMATION: Translocates within the plant. No soil activity. Solubility in water .I ppm. Little or no persistence in the soil. Strong contact activity. Most effective during warm weather. Taken up only by the leaves.

NAMES

OXADIAZON, FORESITE, RONSTAR

3-[2,4-dichloro-5-(1-methylethoxy)phenyl]-5-(1,1-dimethylethyl)
-1,3,4-oxadiazol-2(3H)-one

TYPE: Oxidiazon is an oxadiazole compound used as a selective, preemergence herbicide.

ORIGIN: 1969. Rhone-Poulenc.

TOXICITY: LD_{50} - 3500 mg/kg. Irritating to the eyes.

FORMULATIONS: 50% WP, 2% granules, and 4 Ib flowable.

USES: Ornamentals, and turf. Used on both container and field-grown ornamentals. Used outside the U.S. on rice, orchards, cotton, soybeans, onions and vineyards.

IMPORTANT WEEDS CONTROLLED: Cheatgrass, crabgrass, morningglory, goosegrass, foxtails, barnyardgrass, oxalis, pigweed, sesbania, shepherd 's purse, lambsquarters, bindweed, jimsonweed, smartweed, purslane, and many others.

RATES: Applied at 1-4 Ib actual/A.

APPLICATION: Applied preemergence to the soil surface. Also applied postemergence to weeds in the young seedling stage. Grasses are generally resistant to postemergence applications. Bindweeds are susceptible to both pre and postemergence applications. When used postemergence, apply as a directed spray.Use on both newly planted or established ornamentals. Do not use on red fescue or bentgrass turf or on dichronda, or centipedegrass.

PRECAUTIONS: Perennial weeds, with the exception of bindweeds, are not controlled. Do not disturb the soil surface by cultivation after treatment. Do not apply to the wet foliage of ornamentals or turf. Do not apply to newly seeded turf for 4 months.

ADDITIONAL INFORMATION: More active on broadleaves than grasses. Orchard crops are tolerant at extremely high rates. Very low solubility in water. Non-volatile. Moderately persistent in the soil, giving season-long control. The half life varies from 3-6 months in the soil. Does not leach in the soil. Preemergence, the chemical is absorbed by the young weeds as they grow upwards through the treated zone. More active in a moist soil than under dry conditions. Kills by contract, postemergence. Rainfall or irrigation is required to activate this material. Has little or no effect on the weeds' root systems. Carnations are tolerant of over-the-top, postemergence application. Compatible with other herbicides.

NAMES

ALLOXYDIM-SODIUM, **ALLUXOL, CLOUT, FERVIN,**
GRASIP, GRASIPAN, GRASPAZ, KUSAGARD,
MONALOX, TRITEX, WEED-OUT, ADS

Sodium salt of methyl 2,2-dimethyl-4,6-dioxo-5[1-[(2-propenyloxy) amino]butylidene] cyclohexanecarboxylate

TYPE: Alloxydim-sodium is an oxime compound used as a systemic, postemergence herbicide.

ORIGIN: 1977. Nippon Soda Co., Ltd., of Japan. Being sold in Europe by Schering AG, BASF and Rhone Poulenc.

TOXICITY: LD_{50} - 2260 mg/kg.

FORMULATIONS: 75% SP.

USES: Outside the U.S. on peas, fruit crops, sugarbeets, soybeans, vegetables, beans, sunflowers, potatoes, rape, cotton, and other broadleaf crops.

IMPORTANT WEEDS CONTROLLED: Barnyardgrass, crabgrass, foxtails, johnsongrass, volunteer grains, wild oats, and other grass species.

RATES: Applied at .5-1 kg/ha. May be applied mixed with a crop oil concentrate to increase the activity.

APPLICATION: Applied postemergence to grasses when the first true leaf appears up to the I -2-leaf tillering stage. For perennials apply at the 4-5-leaf stage. A follow up application may be required.

PRECAUTIONS: Do not treat if rain is expected within 4 hours. Not for sale in the U.S. Nutsedge and poa annual are not controlled. Hydroscopic, so keep tightly sealed. Do not use on cereal grains. Some crop yellowing may occur. Grasses emerging after application are not controlled. Do not apply to dew soaked plants. Do not tank mix with other pesticides.

ADDITIONAL INFORMATION: Safe on all broadleaf crops. Usually used in conjunction with other herbicides for broadleaf control. Relatively safe to fish and wildlife. Life in the soil is very short so there are no residue problems. Weed growth decreases almost immediately after treatment. Discoloration of the weeds occur within 5-10 days. Activity is increased by warm wet weather.

NAMES

QUINCLORAC, BAS 514-H, FACET, IMPACT, DRIVE

3,7-dichloro-8-quinolinecarboxylic acid

TYPE: Quinclorac is a quinoline compound used as a selective preplant, preemergence and postemergence herbicide.

ORIGIN: 1984. BASF.

TOXICITY: LD_{50} - 2610 mg/kg. May cause skin irritation.

FORMULATIONS: 50% WP, 1% granules, flowable and formulated with propanil and other herbicides.

USES: Used outside the U.S. on rice, established turf, grain sorghum, soybeans, corn, rape, and cole crops. Experimentally being tested in the U.S. on rice and turf.

IMPORTANT WEEDS CONTROLLED: Jointvetch, signalgrass, crabgrass, barnyardgrass, sesbania, sicklepod, bedstraw, morningglory, plantain, Russian thistle, nightshade, dandelion, and many others.

RATES: Applied at .125 - 1 Ib. ai/A.

APPLICATION: To rice applied preplant incorporated, preemergence and postemergence up to the mid-tillering stage of barnyardgrass. When applied postemergence stop irrigation 1-2 days prior to treatment and irrigate again 1-3 days after application. Avoid any water run-off for 4-5 days. On turf apply either preemergence or early postemergence to crabgrass. On turf do not water for 24 hours after application but rainfall or irrigation is required in 2-7 days. Will control grasses up to the 3-5 tiller stage and broadleaves up to the 4 leaf stage.

PRECAUTIONS: Organic matter in the soil will decrease the herbicidal activity.

Susceptible crops include flax and potatoes. Tolerant weeds include wild onion and wild garlic. Do not apply to bentgrass, centipedegrass, St. Augustine or dichondra turf.

ADDITIONAL INFORMATION: Uptake is by both roots and foliage but root uptake is the most important . Application to moist soil generally is more directive than to dry soil. Long lasting, giving season long control. May be tank mixed with other herbicides. Low fish toxicity. The use with a surfactant will increase the foliar uptake. May be applied by air.

NAMES

QUINMERAC, NIMBUS, GAVELAN, BAS 518

7-chloro-3-methyl-8-quinoline carbonic acid

TYPE: Quinmerac is a quinoline compound used as a selective pre and postemergence herbicide.

ORIGIN: BASF of Germany 1986.

TOXICITY: LD_{50} 5000 mg/kg.

FORMULATION: 50% WP. Often formulated with other herbicides.

USES: Outside the U.S. on cereals, oilseed rape, sugarbeets, peas, potatoes, and other crops.

IMPORTANT WEEDS CONTROLLED: Galium spp. Veronica spp., wild carrots, spurges and other broadleaf weeds.

RATES: Applied at 500-750 g a.i./ha.

APPLICATIONS: Mostly used in combination with other herbicides. Used as both a preemergence and postemergence treatment.

PRECAUTIONS: Not for sale or use in the U.S.

ADDITIONAL INFORMATION: The half life in the soil is from 50-100 days. Non-volatile. Mostly taken up through the root system. When applied preemergence some of the weeds will emerge before they die. When applied postemergence the treated weeds cease to grow and finally die off.

NAMES

GLYPHOSATE, ACCORD, ARCADE, AZURAL, CLYCEL, HOCKEY, KLEENUP, MUSTER, RANGER, RODEO, ROUNDUP, ROUNDUP RT, SHACKLE, SOLADO, SPASOR, STING, AVAIL, JURY, PROTOCOL, RATTLER, ALPHEE, SWING, LAREDO, RIDDER, GALLUP, GLIDER, TREND, STIRRUP, TENDER, BUGGY, TUMBLEWEED, MIRAGE, SPARK, KUSATOBAN, GLIFONOX, EXPEDITE, HONCHO, PONDMASTER, RONDO, GLYPHOGAN

$$OH-\overset{\overset{\textstyle O}{\|}}{C}-CH_2-NH-CH_2-\overset{\overset{\textstyle O}{\|}}{\underset{\underset{\textstyle OH}{|}}{P}}-OH$$

N-(phosphonomethyl) glycine. (Isopropylamine salt)

TYPE: Glyphosate is a phosphoric acid compound used as a broad-spectrum, postemergence, translocated herbicide.

ORIGIN: 1972. Monsanto Chemical Company.

FORMULATIONS: 4 lb/gal soluble solution, 2.7 lb/gal EC. 72% soluable powder.

TOXICITY: LD_{50} - 4900 mg/kg. May cause eye and skin irritation.

IMPORTANT WEEDS CONTROLLED: Most perennial and annual weeds which include quackgrass, johnsongrass, bermudagrass, bentgrass, thistles, milkweed, catlails, kudzu, bindweed, sedges, horsetails, bahiagrass, kikuyugrass, poison ivy, velvetleaf, pigweed, ragweed, fiddleneck, carpetgrass, wild oats, mustards, sicklepod, lambsquarters, jimsonweed, crabgrass, barnyardgrass, leafy spurge, morningglory, kochia, ryegrass, foxtails, dallisgrass, plantain, bluegrass, smartweed, purslane, mesquite, dock, sesbania, teaweed, nightshade, shattercane, chickweed, cocklebur, and many others.

USES: Non-crop weed control as well as pretillage, preemergence or postharvest for annual and perennial weed control in small grains, alfalfa, beans, peas, soybeans, corn, sorghum, rice, and cotton. Used on orchards, groves, and vineyards as a directed spray; also as a directed spray in plantation crops such as rubber, palms, coconut, cocoa, mangoes, coffee, tea, and bananas. Used on apples, Christmas trees, coconuts, root vegetables, alfalfa, soybeans, grapes, grasses, potatoes, lentils, rice, pineapples, forage grasses, forages legumes, sugar beets, garden vegetables, cotton, peanuts, mangoes, guavas, papayas, beans, peas, tomatoes, eggplant, peppers, small fruits and berries, cucurbits, olives, figs, kiwi, cotton, coffee, almonds, filberts, macadamias, kumquats, asparagus, established turf, grasses grown for seed, cherries, avocados, citrus, pears, pecans, pistachios, and walnuts. Also may be used around potable water. Any ornamental species can be planted following an application. May be used as an aquatic herbicide. Used over the top of cotton prior to harvest as a dessicant. Use to release bermudagrass and bahiagrass on non-crop sites.

RATES: Applied at .5-4 Ib actual/A. Effectiveness of control is reduced if applied in high gallonage.

APPLICATION: Apply to perennial weeds that have at least 4-8 leaves so translocation into the plant can occur. Very early treatment of vegetation may reduce the weed control. Good results have been obtained when treating plants at full maturity. Annual weeds are controlled regardless of growth stage. When used as a directed spray, selectivity to certain crops has been achieved. Repeat applications may be needed on certain deep-rooted perennial weeds at 30-45 day intervals. Annual weeds will continue to germinate throughout the season, so retreatment may be necessary. If area to be treated has been mowed, allow regrowth of 8-10 inches high before treating. May be applied over the top of cotton and soybeans with wick, roller, or wiper-type applicators. Also used in recirculating sprayers. May be applied by air. To aquatics apply only to those weeds above the water as submerged aquatics are not controlled. May be applied by injection or frill application to woody plants.

PRECAUTIONS: Do not use or store in galvanized or unlined steel spray equipment as it will react with the metal causing a highly combustible gas. Do not use more than 30 gal of spray solution/A. Avoid spraying the foliage of any desirable crop . Avoid drift. Rainfall occurring within 6 hours of application will reduce the effectiveness. Tillage prior to treatment will reduce perennial weed control. Delay any cultivation of treated perennial weeds for 3-7 days after treatment. In using around trees, do not spray the green bark. Do not mix with dirty water. Reduced control may occur if the treated foliage is covered with dust. Store above 10°F.

ADDITIONAL INFORMATION: Initial activity is fairly slow after application and may not be observed for several days. Shade conditions will slow down the degree of control, but eventually it will be comparable to sunlight conditions. Low toxicity to fish and wildlife. No preemergence activities so a crop can be planted immediately after application. For best perennial weed control do not disturb the plant by tillage until translocation is complete (about 2 weeks). Visible effects are a gradual wilting and

yellowing of the plant.
RELATED MIXTURES:

1. LANDMASTER BW—A combination of 12.9% glyphosate and 20.6% 2,4-D marketed by Monsanto in fallow and reduced tillage grain systems.

2. FALLOWMASTER—A combination of 16.5% glyphosate and 7% dicamba marketed by Monsanto in fallow and reduced tillage grain systems.

NAMES

SULFOSATE, GLYPHOSATE-TRIMESIUM, TOUCHDOWN, OURAGAN, QUATRO, ACEMAX

$$HO\text{-}CO\text{-}CH_2\text{-}NH\text{-}CH_2\text{-}\underset{\underset{OH}{|}}{\overset{\overset{O}{\|}}{P}}\overset{O^-}{\diagup} (CH_3)_3S^+$$

Trimethylsulfonium carboxymethylaminomethyl phosphonate

TYPE: Sulfosate is a phosphate compound used as a selective, postemergence herbicide.

ORIGIN: ICI Ag Products, 1985.

TOXICITY: LD_{50} 750 mg/kg. May cause eye and skin irritation.

FORMULATION: 6EC.

USES: Used in the U.S. on non-crop areas and on non-bearing tree and vines. Used on many crops outside the U.S.

IMPORTANT WEEDS CONTROLLED: Chickweed, annual bluegrass, foxtails, jimsonweed, kochia, wild oats, spurge, ryegrass, barnyardgrass, lambsquarters, crabgrass, nightshade, smartweed and many others.

RATES: Applied .5-3.75 lb a.i./acre.

APPLICATION: Apply thoroughly when weeds are actively growing. Repeat as necessary. Apply with a surfactant or wetting agent. The smaller the weeds, the easier they are to control.

236

PRECAUTION: Avoid drift. Do not use in galvanized steel or unlined steel container. ADDITIONAL INFORMATION: Requires a 6 hour rainfree period after application. On johnsongrass and bermudagrass it should be allowed to reach the seedhead stage before application. May be tank-mixed with other herbicides.

NAMES

GLUFOSINATE-AMMONIUM, BASTA, BUSTER, FINALE, CHALLANGE, DASH, HAYABUSA, IGNITE, CONQUEST, FINAL

$$NH_4 \left[CH_3{-}\overset{\overset{\displaystyle O}{\|}}{\underset{\underset{\displaystyle O}{|}}{P}}{-}CH_2{-}CH_2{-}\underset{\underset{\displaystyle NH_2}{|}}{CH}{-}C\overset{\displaystyle O}{\underset{\displaystyle OH}{}} \right]$$

monoammonium-2-amino-4-(hydroxymethyl phosphinyl) butanoate

TYPE: Glufosinate ammonium is a phosphinic acid compound used as a broad-spectrum, postemergence herbicide with some systemic activity.

ORIGIN: 1981. Hoechst AG of West Germany.

TOXICITY: LD_{50} - 1620 mg/kg.

FORMULATIONS: 1.67 Ib aqueous solution.

USES: Used outside the U.S. in non-crop areas as a contact, non-selective treatment in orchards and vineyards as a postemergence herbicide, in minimum tillage systems, on fallow land, to control the suckers in grape vines, and as a potato desiccant and cotton defoliant. Used experimentally in the U.S. for many of these uses.

IMPORTANT WEEDS CONTROLLED: Wild oats, brome grasses, crabgrasses, barnyard grass, annual bluegrass, ryegrass, foxtails, lambsquarters, jimsonweed, smartweed, purslane, Russian thistle, nightshade, chickweed, bindweed, curly dock, johnsongrass, nutsedge, and many others.

RATES: Applied at .5-2 kg ai/ha.

APPLICATION: Apply to the foliage of actively growing weeds. A lesser rate is required for small weeds. To desiccate potatoes, apply 7-14 days prior to harvest.

237

PRECAUTIONS: Do not apply if rain is expected within six hours. On perennial weeds, burndown of the foliage will occur, but the plant will regrow from the roots. Avoid drift.

ADDITIONAL INFORMATION: More active in warm temperatures. May be tank mixed with soil-active herbicides. Multiple applications may be necessary where perennial weeds are involved. Relatively non-toxic to fish. No soil activity, so seedlings and weeds not yet emerged will not be damaged. Symptoms of the application on weeds are noticeable within 2-5 days. Faster acting than glyphosate but slower than paraquat. More active on broadleaves than on grasses. No soil residual activity. More active when temperatures are over 50°F. Causes ammonium ions to accumulate which disrupts photosynthesis.

NAMES

FOSAMINE-AMMONIUM, KRENITE

$$CH_3 - CH_2 - O - \overset{\displaystyle O}{\underset{\displaystyle O-NH_4}{\overset{\|}{P}}} - \overset{\displaystyle O}{\overset{\|}{C}} - NH_2$$

[ethylhydrogen (aminocarbonyl) phosphate]-ammonium salt

TYPE: Fosamine-ammonium is a phosphoric acid compound used as a postemergence, growth regulator, contact herbicide.

ORIGIN: 1974. E.l. DuPont de Nemours & Company.

TOXICITY: LD_{50} - 10,125 mg/kg.

FORMULATIONS: 4 Ib water-soluble liquid.

USES: Used to control bud growth in most woody species in non-crop areas. Also used for site preparation in conifers.

IMPORTANT WEEDS CONTROLLED: Blackberry, locust, oaks, pines, sumac, sweet gum, and other brush species.

RATES: Applied at 6-12 Ib actual/50-300 gal of water/A. Use the higher rates on large plants, dense growth, or hard-to-control species.

238

APPLICATION: Apply from full leaf in the spring to first fall coloration. For suppression with minimum bud effect, make an application during leaf expansion in the spring. Only the portion of the plant sprayed will be affected. May be applied by air.

PRECAUTIONS: Do not use on food or feed crops. Avoid drift to desirable tree species. It has failed to produce bud break where off-season temperatures are not low enough to cause prolonged dormancy. If rain occurs within 24 hours, it may decrease the effectiveness. Do not apply to brush in standing water.

ADDITIONAL INFORMATION: Safe to fish and wildlife. This compound either prevents spring bud break or provides growth suppression. Plants treated in the summer or fall will not show any symptoms until the following spring when they will not leaf out. Spring application causes growth retardation with abnormal leaf growth. Not effective when applied to the soil. Phytotoxic symptoms occur when applied to new growth in the spring but not when applied in the summer or fall.

NAMES

BUTAMIFOS, CREMART, TUFLER

0-ethyl-0-(5-methyl-2-nitrophenyl)-sec-butyl phosphorothioamidate

TYPE: Butamifos is an organic phosphate compound used as a selective preemergence herbicide.

ORIGIN: 1975. Sumitomo Chemical Co. of Japan.

TOXICITY: LD_{50} - 845 mg/kg.

FORMULATION: 50% EC. 50 WP.

USES: Presently being used in Japan on beans, vegetables, turf and other crops.

IMPORTANT WEEDS CONTROLLED: Barnyardgrass, annual bluegrass, dock, lambsquarters, chickweed, purslane, pigweed, cleavers, dandelion, plantain, henbit, and many others.

APPLICATION: Applied as a preemergence treatment. Rainfall or irrigation is required to take it into the soil.

RATES: Applied at 1-2.5 Ib a.i./ha.

PRECAUTIONS: Not for sale or use in the U.S. Toxic to fish. Avoid drift.

ADDITIONAL INFORMATION: Water solubility 5.1 ppm. Gives season-long weed control. May be soil incorporated in dry areas.

NAMES

BIALAPHOS, BILANAFOS, HERBIACE

Sodium salt of l-2-amino-4-[(hydroxy) (methyl) phosphinol] butyl-l-alamyl-l-alamine

TYPE: Bialaphos is an organic phosphate compound used as a postemergence contact herbicide.

ORIGIN: 1978. Meiji Seika Kaisha Ltd. of Japan.

TOXICITY: LD_{50}- 268 mg/kg.

FORMULATIONS: 32% liquid.

USES: Used outside the U.S. on apples, citrus, grapes, soybeans, potatoes, vegetable crops, noncrop usages and others.

IMPORTANT CROPS CONTROLLED: A wide range of annual and perennial weeds such as shepherd 's purse, chickweed, pigweed, wild oats, lambsquarters, barnyard grass, plantain, annual bluegrass, purslane, foxtails, cocklebur, horsetail, oxalis, crabgrass, curly dock, johnsongrass, quaekgrass, nutsedge, and others.

RATES: Applied at 1-3 kg a.i./ha.

APPLICATION: Applied to fruit and vegetable crops as a directed spray when weeds are young and growing actively. Used as a overall contact treatment in non crop areas and no-till systems.

PRECAUTIONS: Not for sale or use in the U.S.

ADDITIONAL INFORMATION: A natural product that is produced by fermentation . Fast acting with long lasting effects. Inactivated when it reaches the soil . Acts slower than paraquat but faster than glyphosate. Suppresses the regrowth of weeds longer than paraquat but shorter than glyphosate. Active against all stages of weed growth. No preemergence activity so it is suitable for no till situations. Causes ammonia to accumulate in the treated plant reaching 30-100 times normal, resulting in the death of the plant. Relatively non-loxic to fish.

NAMES

ANILOFOS, **AROZIN, RICO, ANILOGUARD**

S-4-chloro-N-isopropyl carbaniloyl methyl-0, 0-dimethyl phosphorodithioate

TYPE: Anilofos is an organic phosphale compound used as a selective postemergence herbicide.

ORIGIN: 1973. Hoechst AG of West Germany.

TOXICITY: LD_{50} - 472 mg/kg. May cause slight eye and skin irritation.

FORMULATION: 30EC, 1.5 and 2% granules.

USES: Transplant paddy rice.

IMPORTANT WEEDS CONTROLLED: Barnyardgrass, spranglelop, small-flower umbrella plant, sedge, and others.

RATES: Applied at .3 - .45 kg ai/ha.

APPLICATION: Apply to transplant rice when the weed seedlings are in the 2.5 leaf-stage. This is usually 4-12 days after transplanting. Apply to a drained field and flood the field 24 hours later. Granules can be applied into the water.

PRECAUTIONS: Not for sale or use in the U.S. Toxic to fish. Do not use on direct-seeded rice.

ADDITIONAL INFORMATION: Absorbed by the weeds through their roots and to some extent through their leaves. After treatment the weeds quit growing, and turn a dark green, and finally die. Compatible wilh other herbicides that control broadleaf weeds.

NAMES

DICLOFOP-METHYL, HOEGRASS, HOELON, ILLOXAN, MASOLON

methyl, 2-[4-(2,4-dichlorophenoxy) phenoxy] propanoate

TYPE: Diclofop-methyl is a selective, diphenylether compound used as a contact and translocated, preemergence and postemergence grass herbicide.

ORIGIN: 1974. Hoechst AG of Germany.

TOXICITY: LD_{50} - 563 mg/kg. May cause eye and skin irritation.

FORMULATIONS: 3EC, 2.36EC, 19% EC.

IMPORTANT WEEDS CONTROLLED: Barnyardgrass, foxtails, goosegrass, volunteer corn, crabgrass, panicum, ryegrass, wild oats, and others.

USES: Wheat, barley, and bermudagrass turf. Being used outside the U.S. on corn, small grains, rape, flax, mustard, sugar beets, potatoes, alfalfa, sunflowers, peas, and other crops.

RATES: Applied at .7-1.4 kg ai/ha.

242

APPLICATION: Postemergence—Applied when grasses are in the 1-4-leaf stage. The larger the grasses, the higher the rate required. Grasses should be growing actively at the time of application. May be applied by air. To volunteer corn,apply before it is 10 inches tall. May be applied with a crop oil concentrate. Preplant—To control downy brome in winter wheat apply as a preplant soil incorporated treatment. Applied to bermudagrass turf to control goosegrass when it is in the 1-4 leaf stage.

PRECAUTIONS: Perennial grasses are not controlled. Toxic to fish. Do not graze treated areas. Annual bluegrass is tolerant of this material. An interval of 5 days must be maintained between the application of this compound and phenoxy or dicamba herbicides. Do not apply when grassy weeds are past the 4-leaf stage. Cool temperatures and wet soil may result in some injury.

ADDITIONAL INFORMATION: Effective on annual grass species only. No effects have been observed on broadleaves. Symptoms on grasses do not show for several days following application. Moisture must be present for the best results. May be tank mixed with bromoxynil and other herbicides and liquid fertilizers. Stage of weed growth is more important than the number of weeds present.

NAMES

ACLONIFEN, BANDUR, BANDREN, CHALLENGE

2-chloro-6-nitro-3-phenoxyaniline

TYPE: Aclonifen is a diphenyl compound used as a selective preemergence herbicide.

ORIGIN: Celamerck of Germany, 1977. Now marketed by Rhone Poulenc.

TOXICITY: LD_{50} 5000 mg/kg. May cause skin irritation.

FORMULATION: 6 SC.

USES: Outside the U.S. on potatoes, peas, lentils, potatoes, corn, sunflowers and winter wheat.

IMPORTANT WEEDS CONTROLLED: Grasses and some broadleaf weeds.

RATES: Applied at 2100-2800 g a.i./ha.

APPLICATION: Applied as a preemergence treatment. Rainfall or overhead irrigation is required to take it into the soil.

PRECAUTION: Not for sale or use in the U.S. Do not incorporate into the soil.

ADDITIONAL INFORMATION: Water solubility is 2.5 ppm. It is absorbed through the young shoots and eventually through the cotyledon and young leaves.

NAMES

ETHOFUMESATE, KEMIRON, NORTRON, TRAMAT, PROGRASS, NORTRANESE, MURENA

2-ethoxy-2,3-dihydro-3,3-dimethyl-5-benzofuranyl methanesulfonate

TYPE: Ethoflumesate is a benzofuran compound being used as a selective, preemergence and postemergence herbicide.

ORIGIN: 1974. Fisons Agrochemicals. Marketed today by Schering AG and Nor-Am.

TOXICITY: LD_{50} - 6400 mg/kg.

FORMULATIONS: 1.5 lb/gal EC. Formulated with other herbicides.

IMPORTANT WEEDS CONTROLLED: Pigweed, crabgrass, barnyardgrass, blackgrass, purslane, shepherd's purse, smartweed, nightshade, kochia, volunteer grains, lambsquarters, foxtails, wild oats, fescues, and others.

USES: Sugar beets, turf, and perennial grasses grown for seed. Being sold outside the U.S. on a number of crops.

APPLICATION: Used as a preemergence treatment applied at or after planting. Rainrall or overhead irrigation is required to move into the soil . May also be soil incorporated, 1-2 inches deep. Non-volatile. Apply to turf either preemergence or postemergence to the weeds.

RATES: Applied at 1-4 kg active/ha.

PRECAUTIONS: Activity is reduced in dry soils. Does not control mustard. Do not use on muck or peat soils. Do not plant any other crop for 12 months after application. Do not use in water with a temperature below 40°F. Do not let stand in tank overnight. May soften PVC hosing.

ADDITIONAL INFORMATION: Water solubility 110 ppm. Absorbed through the emerging shoots of seedlings. Root uptake is relatively slow. Activity may last for 10 weeks from one application. Low toxicity to fish and wildlife. Moisture is required to activate this material. May be mixed with other herbicides.

NAMES

PRONAMIDE, KERB, RAPIER, CLANEX

3,5-dichloro-N-(l,l-dimethyl-2-propynyl)-benzamide

TYPE: Pronamide is an benzamide compound used as a selective, preemergence and early-postemergence herbicide.

ORIGIN: 1965. Rohm & Haas Company.

TOXICITY: LD_{50} - 5620 mg/kg. Mildly irritating to the eyes and skin.

FORMULATIONS: 50% WP. Formulated with other herbicides.

USES: Alfalfa, apples, cherries, grapes, nectarines, peaches, artichokes, fallowland, pears, peas, plums, prunes, rhubarb, trefoil, crown vetch, sainfoin, lettuce, endive, mint, ornamentals, caneberries, clovers, blueberries, and turf.

IMPORTANT WEEDS CONTROLLED: Volunteer barley, bluegrass, bromes, smart-weed, barnyardgrass, nightshade, purslane, cheat, wild oats, chickweed, quackgrass, ryegrass, mustards, shepherd's purse, London rocket, orchardgrass, foxtail, and others.

RATES: Applied at .5-2 Ib active/A.

APPLICATION: Applied as a preemergence treatment. Rainfall or irrigation is required to take it into the soil. Some early postemergence actively on small weeds. If applied when air temperatures exceed 85°F it should be shallowly soil incorporated or watered into the soil immediately.

PRECAUTIONS: A 3-12 month waiting period is required to follow with some crops. Do not use on high organic matter soils. Not effective against established johnsongrass, nutgrass, composites, sesbania, puncturevine, dandelions, pineapple weed, or legumes. Do not use on dichondra, bluegrass, ryegrass, fescues, or bentgrass turf. Do not use on golf greens.

ADDITIONAL INFORMATION: More effective in cool weather. More effective preemergence than postemergence. Little foliar activity since it requires root uptake. May be applied by air. Absorption by the roots is necessary, so it may take 3-5 weeks to control poa annual. Controls poa annua at all stages of growth. As soil temperatures increase the product will degrade faster. May be used with other herbicides. Moisture is essential to move the herbicide into the soil.

NAMES

ISOXABEN, CENT-7, FLEXIDOR, X-PAND, RATIO GALLERY, ELSET, KNOCK-OUT, SEXTAN

N-[3-(1-ethyl-1-methylpropyl)-5-isoxazolyl]-2,6-dimethoxybenzamide

TYPE: Isoxaben is a amide compound used as a selective preemergence herbicide.

ORIGIN: 1982. Elanco Products.Co. Now marketed by DowElanco.

TOXICITY: LD_{50} -10,000 mg/kg. May cause slight eye and skin irritation.

FORMULATION: 50% SC. 75% DF. Formulated with other herbicides.

USES: Turf and ornamentals in the U.S. Used on small grains, fruit trees, vineyards and other crops outside the U.S.

IMPORTANT WEEDS CONTROLLED: Shepherd's purse, field bindweed, poppy, buttercup, mayweed, groundsel, corn spurry, chickweed, pennycress, wild radish, and many other broadleaf weeds.

RATES: Applied at 50-300 g ai/ha.

APPLICATION: Applied either pre or postemergence to cereals, but preemergence to the broadleaf weeds. Rainfall or irrigation within 21 days will carry it into the soil. Used preemergence on other crops.

PRECAUTIONS: Some rotational crops may be injured if planted immediately following the treated crop. Grasses are not controlled. Do not apply to ajuga, candy tuft, sediums or spurge ornamentals. Do not use on new planted or transplanted ornamentals.

ADDITIONAL INFORMATION: Applied in the fall season, control of spring germinating broadleaf weeds can be expected. Low water solubility so it does not leach in the soil. May be combined with other herbicides. Stable on the soil surface for up to 21 days.

RELATED MIXTURES:

1. SNAPSHOT DF—A combination of isoxaben and oxyzalin marketed by DowElanco as an ornamental herbicide.

2. SNAPSHOT TG—A combination of isoxaben and trifluralin marketed by DowElanco as a turf, Christmas tree, non crop and ornamental herbicide.

NAMES

FOMESAFEN, FLEX, REFLEX, DARDO

5-[2-chloro-4-(trifluoromethyl)phenoxy]N-(methylsulfonyl)-2-nitrobenzamide

TYPE: Fomesafen is a diphenyl compound used as a postemergence contact herbicide with soil residual activity.

ORIGIN: 1977. ICI-Plant Protection Div. of England.

TOXICITY: LD_{50} - 1250 mg/kg. May cause eye and skin irritation.

FORMULATION: 2 LC.

USES: Soybeans. Being used outside the U.S. on soybeans and other crops.

IMPORTANT WEEDS CONTROLLED: Morningglory, cocklebur, nightshade, sicklepod, ragweed, sesbania, jimsonweed, lambsquarters, pigweed, teaweed, velvetleaf, yellow nutsedge, and others.

RATES: Applied at .25 - .375 ai/A.

APPLICATIONS: Applied as a postemergence broadcast application when the weeds are 1-3 inches tall, usually 14-21 days after planting soybeans.. Apply with a non-ionic surfactant or a crop oil concentrate.

PRECAUTIONS: Grasses are not controlled. Apply before soybeans bloom. Rainfall within 4 hours of application may decrease the control.

ADDITIONAL INFORMATION: May be tank mixed with other herbicides for grass control. Up to 4 weeks soil residual weed control can be obtained from a postemergence application. Rainfall must occur within 5 days if soil activity is to be maintained. Day temperatures should be above 70°F and soil temperatures above 60°F for the best results.

248

NAMES

BENTAZON, BASAGRAN, TROPHY, ADAGIO, DEPEND, PLEDGE, LEADER, BASAMAIS

(3-(1-methylethyl)-1H,-2,1,3-benzothiadiazin-4(3H)-one 2,2-dioxide)

TYPE: Bentazon is a benzothiadiazole compound used as a selective, postemergence herbicide.

ORIGIN: 1968. BASF of Germany.

FORMULATIONS: 4 Ib/gal water-soluble sodium salt.

TOXICITY: LD_{50} - 1100 mg/kg. May cause eye and skin irritation.

USES Soybeans, turf, beans, mint, peanuts, sorghum, corn, rice, and peas. Registered on a number of crops in different countries.

IMPORTANT WEEDS CONTROLLED: Velvetleaf, nutsedge, Canada thistle, morningglory, ragweed, cleavers, mustard, mayweeds, jimsonweed, sunflower, teaweed, mallow, smartweed, cocklebur, hairy nightshade, and many others.

RATES: Applied at .5-2 Ib ai/A.

APPLICATION: Applied as a postemergence treatment to soybeans which are tolerant at all stages of growth. Broadleaf weeds in the 2-10-leaf stage are the most readily controlled. May be applied by air. Weeds should be growing actively. Dry beans should have the first trifoliate leaves fully extended and peas should have three pairs of leaves before application, or injury will result. To rice, apply only to weeds emerged above the water Ievel.

PRECAUTIONS: Do not apply to turf until it is well established. Do not apply to blackeyes or garbanzo beans. Rain within 8 hours will reduce the effectiveness. Sicklepod, beggerweed, horsenettle, dandelion, and vetches are among the resistant weeds, as are grasses. Do not apply to soybeans growing under stress conditions, such as flooding, as injury may result. Prolonged periods of cold weather will give poor results. Black nightshade is not controlled.

ADDITIONAL INFORMATION: No preemergence activity has been noted. Translocated within the plant. Cocklebur is an extremely susceptible weed, and can be controlled when 12-18 inches tall. The addition of a wetting agent may increase the activity of this material. May be combined with other herbicides. Most grain crops and large-seeded legumes show considerable tolerance to this material. Water solubility of 570 ppm. The higher the temperature, the more effective this material becomes. The soil type does not affect the application rate. Results can be seen 2-7 days after application. Nutgrass can be controlled with a split application. More active in warm weather. May be applied with 28% nilrogen solution to increase velvetleaf control in soybeans.

RELATED MIXTURES:

1. LADDOK, PROMPT—A combination of bentazon and atrazine marketed by BASF. Laddok is marketed on corn and sorghum, while Prompt is used on certain turf species.

NAMES

DIETHATYL-ETHYL, ANTOR

N-chloroacetyl-N-(2,6-diethylphenyl)-glycine ethyl ester

TYPE: Diethatyl-ethyl is a chloroacetanilide compound used as a selective, preemergence herbicide.

ORIGIN: 1974. Hercules, Inc. (now Nor-Am Chemical Co.)

TOXICITY: LD_{50} - 2300 mg/kg. May cause irritation to eyes and skin.

FORMULATIONS: 4 lb/gal EC.

USES: Sugar beets, red beets, spincah and bermudagrass grown for seed. Being used on sunflowers, potatoes, peanuts, beans, and broccoli and other crops in other countries.

IMPORTANT WEEDS CONTROLLED: Barnyardgrass, blackgrass, canarygrass, wild oats, crabgrass, panicum, foxtails, groundcherry, nightshade, pigweed, ryegrass, shepherd's purse, sowthistle, and others.

RATES: Applied at 1.7-6.7 kg/ha (1 1/2-6 Ib/A.)

APPLICATION: May be applied preemergence or preplant incorporated to a depth of about 1-2 inches. Rainfall is required to move it into the soil.

PRECAUTIONS: Soil incorporation is necessary in dry areas. Do not allow the spray mixture to set in the spray tank over night. Do not use on muck soils. Toxic to fish.

ADDITIONAL INFORMATION: More active on grasses than broadleaves. Water solubility 105 ppm. Absorbed by the emerging shoots as they grow through the treated soil. Very limited postemergence activity. Non-volatile, but moisture is needed for activation. May be tank mixed with other herbicides. May be used with liquid fertilizers.

NAMES

FLURIDONE, BRAKE, SONAR, COMPEL, PRIDE

1-methyl-3-phenyl-5-(3-trifluoromethyl)
phenyl)4(1H) pyridinone

TYPE: Fluridone is a pyridinone compound used as a postemergence, selective herbicide.

ORIGIN: 1975. Elanco Products Co. Now marketed by DowElanco.

TOXICITY: LD$_{50}$- 10,000mg/kg. May cause slight eye irritation.
FORMULATIONS: 4 AS, 5% pellet.

USES: An aquatic herbicide in ponds, lakes, reservoirs, drainage canals and irrigation canals.

IMPORTANT WEEDS CONTROLLED: Aquatic weeds such as coontail, water milfoil, elodea, hydrilla, bladderwort, naiad, pond weeds, water lily, duckweed and others.

RATES: To ponds use at .25-4 Ibs. ai/surface acre depending on water type and depth.

APPLICATION: Applied at any stage of weed growth when they are actively growing.

PRECAUTIONS: Algae is not controlled. Do not treat in narrow strips. Do not use for irrigation water for 7-30 days depending upon the crop.

ADDITIONAL INFORMATION: Slow action prevents oxygen depletion. Absorbed by aquatic weeds through the water or through their root systems. Low toxicity to fish. Cotton exhibits a true physiological tolerance to this material. Other species in the Malvacea family of plants are severely injured. Water solubility 12 ppm. Effective at extremely low rates. The half-line in pond water is 14 days or less. In aquatics, little or no weed control is noted for 7-10 days and may take up to 90 days. Inhibits the formation of carotene.

NAMES

SETHOXYDIM, CHECKMATE, EXPAND, FERVINAL, GRASIDIM, NABU, NABUGRAM, POAST, POAST PLUS, VANTAGE, ALJADEN

2-[1-(ethoxyimino)butyl-5-[2-(ethylthio) propyl]-3-hydroxy-2-cyclohexen-1-one]

TYPE: Sethoxydim is a oxime compound used as a selective, postemergence herbicide.

ORIGIN: 1978. Nippon Soda of Japan. Being marketed in the U.S. and some other counries by BASF.

TOXICITY: LD_{50} - 2676 mg/kg. Causes eye and skin irritation.

FORMULATION: 1 and 1.5 lb/gal EC.

USES: Soybeans, cotton, potatoes, alfalfa, fruiting vegetables, peanuts, strawberries, sunflowers, peas, beans, apples, pears, grapes, citrus, blueberries, corn, onions, garlic, tobacco seedbeds, sweet potatoes, sugarbeets, flax, artichokes, caneberries, cucurbits, Brussels sprouts, cauliflower, broccoli, cabbage, lettuce, celery, spinach and ornamentals. Used in other countries on many broadleaf crops to control grasses.

IMPORTANT WEEDS CONTROLLED: All annual grasses except Poa spp. and most perennial grasses. Nutsedge is not controlled.

RATES: Applied at .1 -1 lb ai/A.

APPLICATION: Applied as a postemergence treatment. Treat annual grasses up to the 6-8-leaf stage. Treat quackgrass at the 2-3-leaf stage or 6-8 inches high. Treat johnsongrass at 5-7-leaf stage or 15-18 inches high. A split application may be used. Apply with a non-phytotoxic crop oil concentrate.

PRECAUTIONS: Does not control broadleaf weeds or sedges. All grass crops are sensitive to this product. Not too effective when temperatures drop below 60°F. Do not apply if rainfall is expected within one hour.

ADDITIONAL INFORMATION: All broad leaf plants and sedges are tolerant to this product. Persistence in the soil is very short, and preemergence activity is low. Absorbed rapidly through the foliage. Noticeable control of perennials the following season has been observed. Some fescue species are tolerant. May take up to 3 weeks for complete control. Control symptoms will be noticed in 5-7 days. May be tank mixed with other herbicides. The addition of a crop oil concentrate increases the activity. May be applied by air.

NAMES

TRALKOXYDIM, PP604, GRASP, SPLENDOR, ACHIEVE

2-[1-(ethoxyimino) propyl]-3-hydroxy-5-(2,4,6-trimethylphenyl) cyclohex-2-inone

TYPE: Tralkoxydim is a cyclohexane compound used as a selective postemergence herbicide.

ORIGIN: ICI Ag Products 1987.

TOXICITY: LD_{50} 934 mg/kg.

FORMULATIONS: 1EC, 2.5 SC.

USES: Outside the U.S. on cereals.

IMPORTANT WEEDS CONTROLLED: Wild oats, ryegrass, foxtails, canarygrass, blackgrass and others.

RATES: Applied at 150-350 g a.i./ha.

APPLICATION: Applied as a postemergence treatment when the weeds are in the 2 leaf stage to the second node shape. Apply with a nonionic surfactant for the best results.

PRECAUTIONS: Not for sale or use in the U.S. Weeds must be growing vigorously to be controlled.

ADDITIONAL INFORMATION: May be tank-mixed with other herbicides. Rain within 15 minutes of the application will not effect the weed control. Enters the foliage and moves rapidly in the phloem tissue to the growing sprouts. May be used on winter varieties of wheat, barley and rye and on durum wheat. Crop tolerance doesn't seem to vary by variety.

NAMES

CLETHODIM, SELECT

(E)-2-[1-1[(3-chloro-2-propenyl)oxy]imino]propyl]5-[2-(ethylthio)propyl]-3-hydroxy-2-cyclohexen-l-one

TYPE: Clethodim is an oxime compound used as a selective postemergence herbicide.

ORIGIN: 1985. Chevron Chemical Co. Now being marketed by Valent in the U.S. and other companies outside the U.S.

TOXICITY: LD_{50}- 1360mg/kg.

FORMULATION: 2EC.

USES: Cotton and soybeans. Experimentally being tested on peanuts, sugar beets potatoes, beans, alfalfa, vegetable crops and others. Used on these crops outside the U.S. IMPORTANT WEEDS CONTROLLED: Annual grasses, seedling johnsongrass and volunteer cereals.

RATES: Applied at .075-.25 kg a.i./ha.

APPLICATION: On annual grasses, apply when they are 2-24 inches tall. On perennial grasses, apply when they are young and repeated applications may be necessary. Use with a crop oil concentrate.

PRECAUTIONS: Broadleaves are not controlled. Avoid drift. Do not apply a broadleaf herbicide within 1 day of application. Rainfall within 1 hour of application may reduce the effectiveness.

ADDITIONAL INFORMATION: Use with a crop oil concentrate when applied postemergence. Slow acting taking 7-14 days for control. Do not apply to plants in a stressed condition.

NAMES

CYCLOXYDIM, FOCUS, LASER, STRATOS, BAS-517

2-1 1-(ethoxyimino)butyl]-3-hydroxy-5-(3-thiayl)-2-cyclohexen- 1 -one

TYPE: Cycloxydim is an oxime compound used as a postemergence selective herbicide.

ORIGIN: 1983. BASF.

TOXICITY: LD_{50} 3940 mg/kg. May cause slight skin irriation.

FORMULATION: 2 EC.

USES: Outside the U.S. on broadleaf crops such as alfalfa, beans, peas, beans, cole crops, ornamentals, cucifers, celery, cocoa, coffee, citrus, cotton, peanuts, potatoes, rape, soybeans, sugar beets, sunflowers, vegetable crops and others.

IMPORTANT WEEDS CONTROLLED: Quackgrass, wild oats, bromes, bermudagrass, sandbur, crabgrass, barnyardgrass, ryegrass, panicum, volunteer grains, foxtails, johnsongrass and many others.

RATES: Applied at 200-500 g a.i./ha.

APPLICATION: Applied to annual grasses when they are in the 2-4 leaf stage. To bermudagrass, apply when the stolons are up to 10 inches in height. To johnsongrass, apply before it is 24 inches tall. To quackgrass, apply when it is 4-8 inches tall. A second application may be required with perennial grasses. Use with a crop oil concentrate for the best results. May be applied by air.

PRECAUTIONS: Not for sale or use in the U.S. Toxic to fish. Broadleaf weeds are not controlled. Herbicidal activity may be decreased if mixed with broadleaf herbicides.

ADDITIONAL INFORMATION: Absorbed rapidly through the foliage. Translocates both upward and downward in the grass plants. Rainfall one hour after application will not decrease the activity. Sedges, red fescue and annual bluegrass appear to be tolerant of this product. Control symptoms will be noticed in 3-4 days with complete kill in about 3 weeks. Some preemergence activity. More active at higher temperatures.

NAMES

METHAZOLE, MEZOPUR, PAXILON, PROBE, TUNIC, BIOXONE

2-(3,4'-dichlorophenyl)-4-methyl-1,2,4oxadiazodidine-3,5-dione

TYPE: Methazole is an phenyl urea compound used as a selective pre and postemergence herbicide.

ORIGIN: 1969. Velsicol Chemical Co. Being sold today by Sandoz Agro.

TOXICITY: LD_{50} - 2500 mg/kg. Causes eye and skin irritation.

FORMULATION: 75% WDG.

USES: Cotton in the U.S. Used outside the U.S. on onions, potatoes, citrus, alfalfa, tea, orchards, grapes and other crops.

IMPORTANT WEEDS CONTROLLED: Pigweed, lambsquarters, purslane, carpetweed, mustard, morningglory, ragweed, cocklebur and other.

RATES: Applied al 1-425 kg a.i./ha.

APPLICATIONS: Applied either preemergence or postemergence as a directed spray when the cotton is 3-6 inches tall or taller. Apply to moist soil if possible. On other crops apply preemergence or as a directed spray.

PRECAUTIONS: Do not use on soils above 4% organic matter. Toxic to fish. Do not soil incorporate.

ADDITIONAL INFORMATION: Slow to leach from the soil. Solubility in water is 1.5 ppm. May be mixed with other herbicides.

NAMES

FLUOROCHLORIDONE, R-40244, RACER, RAINBOW

3-chloro-4-(chloromethyl)-1-(3-trifluoromethyl)phenyl)-2-pyrrolidinone

TYPE: Fluorochloridone is a pyrrolidione compound used as a selective preemergence herbicide.

ORIGIN: 1982. Stauffer Chemical Co. Being marketed by ICI.

TOXICITY: LD_{50} - 1820 mg/kg. May cause eye and skin irritation.

FORMULATION: 2EC. 4EC, 25CS.

USES: Outside the U.S. on carrots, potatoes, sunflowers, cereals, corn, cotton, lentils, peas, wheat, sugarcane, trees, plantation crops, vines and non-crop usages.

IMPORTANT WEEDS CONTROLLED: Annual bluegrass, barnyardgrass, crabgrass, foxtails, nightshade, chickweed, galinsoga, groundcherry, henbit, kochia, lambsquarters, mallow, pigweed, ragweed, purslane, Russian thistle, shepherd's purse, spurge, teaweed, velvetleaf, mustard and others.

RATES: Applied at .375-1 kg a.i./ha.

APPLICATION: Applied as a surface applied preemergence herbicide. Light incorporation into the soil surface is used under dry conditions.

PRECAUTIONS: Not for sale or use in the U.S. Avoid drift. Somewhat toxic to fish. Do not let stand overnight in the spray tank. Sensitive crops include tomatoes, cole crop, cucurbits. May persist in the soil.

ADDITIONAL INFORMATION: May be combined with other herbicides to give a broader spectrum of control. Water solubility is 28 ppm. Appears to inhibit carotenoid synthesis in plants so treated plants become bleached. More effective on broadleaves than grasses. Use higher rates when used for total vegetation control in non-crop areas.

NAMES

CLOMAZONE, COMMAND, GAMIT, MAGISTER, COLZOR

2-(2-chlorophenyl) methyl-4,4-dimethyl-3-isoxazolidinone

TYPE: Clomazone is an isoxazolidinone compound used as a selective preemergence or preplant incorporated herbicide.

ORIGIN: 1982. FMC Corp.

TOXICITY: LD_{50} - 1369 mg/kg. May cause some eye or skin irritation.

FORMULATION: 4EC.

USES: Soybeans, pumpkins and peas. Also used on fallow land to which wheat will be planted. Experimentally being tested on potatoes, tobacco, cotton, dry beans, peanuts, cucurbits, sweet potatoes and other crops. Used outside the U.S. on rape, sugarcane and tobacco.

IMPORTANT WEEDS CONTROLLED: Crabgrass, barnyardgrass, panicum, foxtails, johnsongrass (seedlings), velvetleaf, pigweed, ragweed, lambsquarters, jimsonweed, smartweed, purslane, prickly sida, nightshade, cocklebur and others.

RATES: Applied at .75 -1.25 lb ai/A.

APPLICATION: Applied as either a preemergence or preplant incorporated treatment. May be tank-mixed with other herbicides to increase the spectrum of control. Incorporate within 3 hours of application.

PRECAUTIONS: Weeds not controlled include, carpetweed, giant ragweed, hemp sesbania, morningglory, sedges, sicklepod, mustard and sunflower. Foliar contact or vapors may cause visual symptoms of chlorosis to nearby sensitive plants. Do not plant small grains or alfalfa in the fall of the year of application or in the spring of the following year. Avoid drift. May persist in the soil.

ADDITIONAL INFORMATION: Relatively non-toxic to fish. Absorbed by the roots and shoots of the weed. Susceptible weeds emerge from the soil, but are devoid of pigmentation so plant death occurs within a short period of time . Half life in lhe soil is 15-45 days. May be used with liquid fertilizers.

RELATED MIXTURES:

1. COMMENCE—A combination of trifluralin and clomazone developed by FMC and DowElanco for usage on soybeans as a preplant incorporated treatment.

NAMES

SULFENTRAZONE, F-6285

[1-[2,4-dichloro-5-[N-(methylsulfonyl) amino]-phenyl]-1,4-
dihydro-3-methyl-4-(difluoromethyl)-5H-triazol-5-one]

TYPE: Sulfentrazone is a triazolone compound used as a selective preplant incorporated on preemergence herbicide.

ORIGIN: FMC Corp., 1989.

TOXICITY: LD_{50} 2000 mg/kg. May cause slight skin irritation.

FORMULATION: 4 lb/gal flowable.

USES: Experimentally being tested on sugarcane, soybeans, peas and other crops.

IMPORTANT WEEDS CONTROLLED: Barnyardgrass, beggarweed, cocklebur, crab-grass, jimsonweed, lambsquarters, nightshade, nutsedge, pigweed, prickly sida, smart-weed, and many others.

RATES: Applied at .38-.5 lb a.i./acre.

APPLICATION: Applied as either a preplant incorporated, or as a preemergence treatment. Requires soil moisture or rainfall to activate the material.

PRECAUTION: Used on an experimental basis only. Cotton and sugarbeets in rotation may be injured. Weeds not controlled include foxtails, sesbania, shattercane and sicklepod.

ADDITIONAL INFORMATION: Gives season long control. Preplant incorporate under dry conditions. Greater activity is observed in light shady soils. Soil pH does not affect the activity. Plants emerging from treated soils will turn necrotic and die shortly after exposure to light. Taken up by both the roots and foliage. Moderately mobile in the soil. Sunlight does not break it down. Non volatile.

NAMES

PYRAZOXYFEN, PAICER, SL-49

2-[4-(2,4-dichlorobenzoyl)-1,3-dimethylpyrazol-5-yloxyl]acetophenon

TYPE: Pyrazoxyren is an acetophenone compound used as a selective pre and postemergence herbicide.

ORIGIN: 1984. Ishihara Sangyo Kaisha Ltd. of Japan.

TOXICITY: LD_{50}- 1644 mg/kg.

FORMULATION: 10% granules. Formulated with other compounds.

USES: Paddy rice.

IMPORTANT WEEDS CONTROLLED: Barnyardgrass, sedges, bulrush and other annual and perennial weeds.

RATES: Applied at 3 kg. ai/ha.

APPLICATION: Applied as either a preemergence or postemergence soil treatment generally 1-7 days after transplanting of the rice. Flood the fields after treatment. Can be used on direct seeded rice but only when the temperatures are above 35°C.

PRECAUTIONS: Not for sale or use in the U.S. Toxic to fish. Do not use on upland rice as the activity is extremely reduced unless most of the weeds are submerged.

ADDITIONAL INFORMATION: Broad spectrum control of both annual and perennial

weeks. Excellent rice tolerance. Control can be expected for 3-5 weeks. Low temperatures result in reduced control. Symptoms do not appear on the weeds until 4-5 days after
treatment and total kill takes 1-2 weeks.

NAMES

PYRAZOLYNATE, SANBIRD

4-(2,4-dichlorobenzoy)-1,3 dimethylpyrazol-5-
yl-toluene-4-sulfonate

TYPE: Pyrazolynate is a pyrazole compound used as a selective preemergence herbicide.

ORIGIN: Sankyo of Japan, 1990.

TOXICITY: LD_{50} 9550 mg/kg.

FORMULATION: 10% granules. Formulated with other herbicides.

USES: Outside the U.S. on paddy rice.

IMPORTANT WEEDS CONTROLLED: Many grass and sedge species.

RATES: Applied at 3 kg a.i./ha.

APPLICATION: Applied prior to seeding or transplanting.

PRECAUTION: Not for sale or use in the U.S.

NAMES

BENFURESATE, CYPERAL, MORLENE

CH₃ — CH₂ — S(=O)(=O) — O — [benzofuran structure] with CH₃, CH₃ groups

$$CH_3-CH_2-\overset{\displaystyle O}{\underset{\displaystyle O}{\overset{\|}{\underset{\|}{S}}}}-O-\text{(2,3-dihydro-3,3-dimethylbenzofuran-5-yl)}$$

2,3-dihydro-3,3-dimethyl-5-benzofuranyl ethanesulphonate

TYPE: Benfuresate is a benzofuran compound used as a selective pre and postemergence herbicide.

ORIGIN: 1976. Fisons Lld. Now it is being marketed by Schering AG of West Germany.

TOXICITY: LD_{50} 2031 mg/kg. May cause skin irritation.

FORMULATION: 4 EC.

USES: Used outside the U.S. on cotton, corn, potatoes, rice, peas, beans, fruit trees and other crops.

IMPORTANT WEEDS CONTROLLED: Nutsedge (Cyperus spp.) as well as crabgrass, barnyardgrass, goosegrass, panicums, foxtails, johnsongrass, pigweed, lambsquarters, spurge, purslane, chickweed, morningglory, smartweed, teaweed, nightshade and others.

RATES: Applied at .5-2.8 kg. ai/ha.

APPLICATION: Used as a preplant incorporated preemergence and postemergence herbicide. Has some activity postemergence against nutsedge.

PRECAUTION: Not for sale or use in the U.S. Susceptible crops to a preemergence application include sugarbeets, crucifer crops, carrots, cereals, cucurbits, lentils, lettuce, rape, peanuts, sesame, sorghum, soybeans, tomatoes and others. Poor postemergence activity has been shown on all weeds, except nutsedge. Resistant weeds include velvetleaf, ragweed, wild oats, jimsonweed, kochia, wild buckwheat, cocklebur and others.

ADDITIONAL INFORMATION: Nutsedge is one of the most susceptible weed species to this compound. Controls plants by the emerging shoots as it grows through the treated soil layer. May be mixed with other herbicides to increase its effectiveness. Cotton is considered to be one of the most tolerant crops to the product.

NAMES

FLURTAMONE, BENCHMARK, RE 40885

5-(methylamino)-2-phenyl-4-(3-trifuoromethylphenyl)3-(2H)-furanone

TYPE: Flurtamone is a furanone compound used as a preplant or preemergence selective herbicide.

ORIGIN: 1986. Chevron Chemical Co. Being developed by Rhone Poulenc.

TOXICITY: LD_{50} - 500 mg/kg. May cause eye and skin irritation.

FORMULATION: 50% WP.

IMPORTANT WEEDS CONTROLLED: Controlls a number of broadleaf weeds and suppresses certain grass species.

RATES: Applied at .25-1 Ib a.i./ha.

USES: Experimentally being tested on cotton, peanuts, sorghum, safflower, barley, peas, sunflowers, tree and vine crops, and others.

APPLICATION: Used as a soil incorporated preplant treatment or applied preemergence or early postemergence.

PRECAUTIONS: Used on an experimental basis only.

ADDITIONAL INFORMATION: More effective on broadleaves than grasses.

NAMES

PYRITHIOBAC, STAPLE, DPX-PE 350, KIH-2031

Sodium 2-chloro-6-(4,6-dimethoxyprimidin-2-ylthio) benzoate

TYPE: Pyrithiobac is a benzoate compound used as a selective postemergence herbicide.

ORIGIN: 1989 Kumiai Chemical Industry of Japan. Being developed in the U.S. by DuPont.

TOXICITY: LD_{50} 1000 mg/kg.

USES: Experimentally being tested on cotton.

IMPORTANT WEEDS CONTROLLED: Velvetleaf, pigweed, coffeeweed, morningglory, kochia, smartweed, teaweed, johnsongrass, cocklebur and others.

RATES: Applied at 70 g a.i./ha.

APPLICATION: Applied postemergence to the weeds when they are young and actively growing. Use with a crop oil concentrate.

PRECAUTIONS: Used on an experimental basis only.

ADDITIONAL INFORMATION: Temporary yellowing may occur if applied when the cotton is stressed.

NAMES

FLUMIPROPYN, S-23121

**(R,S)-N-[4-chloro-2-fluoro-5-[(1-methyl-2-propynyl) oxy] phenyl]-
3,4,5,6-tetrahydrophthalimide**

TYPE: Flumipropyn is a phthalimide compound used as a selective pre and postemergence herbicide.

ORIGIN: Sumitomo Chemical Co. of Japan 1987. Being developed in the U.S. by Valent.

TOXICITY: LD_{50} 5000 mg/kg. May cause eye irritation.

FORMULATIONS: 10% SC.

USES: Experimentally being used on cereals.

IMPORTANT WEEDS CONTROLLED: Veronica spp, Galium spp, Viola spp, velvetleaf, morningglory, cocklebur and many other broadleaf weeds.

RATES: Applied as a preemergence treatment or as a postemergence treatment when the weeds are young. May be used in combinations with other herbicides. Apply when the crops and weeds are in the 1-4 leaf stage.

PRECAUTIONS: Used on an experimental basis only. Grasses, with the exception of annual bluegrass, are not controlled.

ADDITIONAL INFORMATION: A contact herbicide that is readily absorbed into the plant tissue. Fast acting. Controls many hard to control broadleaf weeds.

NAMES

SULCOTRIONE, ICIA-0051, GALLEON

2-(2-chloro-4-mesylbenzoyl) cyclohexane-1,3-dione

TYPE: Sulcotrione is a triketone compound used as a selective postemergence herbicide.

ORIGIN: 1990 ICI Ag Products.

TOXICITY: Under investigation.

FORMULATION: 300 SP

USES: Experimentally on corn and sugarcane.

IMPORTANT WEEDS CONTROLLED: Crabgrass, barnyardgrass, panicum, nightshade, lambsquarters, smartweed and others.

RATES: Applied at 300-450 g a.i./ha.

APPLICATION: Applied as a postemergence treatment.

PRECAUTION: Used on an experimental basis only.

ADDITIONAL INFORMATION: Used as an alternative to atrazine. Cross resistance with triazine is considered unlikely. Some soil residual.

FLUOROGLYCOFEN-ETHYL, COMPETE, RH-0265, SATIS, SIMITAR

2-ethoxy-2-oxoethyl-5-[2-chloro-4-(trifluoromethyl)
phenoxy]-2-nitrobenzoate

TYPE: Fluoroglycofen-ethyl is a benzoate compound used as a selective postemergence herbicide.

ORIGIN: Rohm & Haas, 1990.

TOXICITY: LD_{50} 1500 mg/kg. May cause slight eye and skin irritation.

FORMULATION: 2EC, 5 & 20% WP.

USES: Experimentally being used on wheat and barley.

IMPORTANT WEEDS CONTROLLED: Galium spp, Viola spp, Veronica spp, and other broadleaves are controlled.

RATES: Applied at 10-80 g a.i./ha.

APPLICATION: Applied as a postemergence treatment when weeds are 2-8 cm in size and the cereals are between 3 leaves and stem elongation.

PRECAUTION: Used on an experimental basis only. Grasses or perennial weeds are not controlled. Under some conditions, slight injury to cereals may occur in the form of small white spots on the leaves. These will disappear in a few weeks. Toxic to fish.

ADDITIONAL INFORMATION: May be mixed with other herbicides.

NAMES

UBI-C4243, UBI-9086

Chemistry has not been released

ORIGIN: Uniroyal, 1990.

TOXICITY: LD_{50} 5000 mg/kg.

FORMULATION: 10% EC.

USES: Experimentally being tested as a preemergence herbicide on wheat, corn, peas, lentils, sugarcane, sorghum and chemical follow. Also as a post hervest desiccant on potatoes, soybeans, drybeans, cotton, rape, sunflower, alfalfa and other crops.

IMPORTANT WEEDS CONTROLLED: Pigweed, purslane, lambsquarters, teaweed, ragweed, velvetleaf, jimsonweed, kochia, chickweed, henbit, shepardspurse, nightshade, ryegrass, foxtails, goosegrass and others.

RATES: Applied at .017-.14 kg a.i./ha.

APPLICATION: Applied as a preemergence or postemergence herbicide treatment or as a post harvest desiccant.

PRECAUTIONS: For use on an experimental basis only. Avoid drift. Rainfall is required to activate this material.

ADDITIONAL INFORMATION: May be combined with other herbicides. A soil persistant herbicide. Causes desiccation of plant tissue when exposed to light.

GLOSSARY

Glossary

SPRAYER CALIBRATION

Calibration of your spraying equipment is very important. It should be done at least every other day of operation to insure application of the proper dosages. This is probably the most important step in your whole spraying operation since applying incorrect amounts may do much more damage than good.

If a lower rate is desired it may be obtained by increasing the speed, reducing the speed, increasing the pressure or changing to a larger nozzle or a combination of the three.

GENERAL CALIBRATION

I. Method I.
   A. Measure out 660 feet.
   B. Determine the amount of spray put out in traveling this distance at the desired speed.
   C. Use this formula:

$$\text{gallons/acre} = \frac{\text{gallons used in 660 feet x 66}}{\text{swath width in feet}}$$

   *D. Fill the tank with the desired concentration.

II. Method II.
   A. Fill spray tank and spray a specified number of feet.
   B. After spraying refill tank measuring the quantity of material needed for refilling.
   C. Use this formula:

$$\text{gallons/acre} = \frac{43560 \text{ x gallons delivered}}{\text{swath length (ft.) x swath width (ft.)}}$$

   *D. Fill the tank with the desired concentrate.

III. Method III.
   A. Measure 163 feet in the field.
   B. Time tractor in 163 feet. Make two passes to check accuracy.
   C. At edge of field adust pressure valve until you catch 2 pints (32 ounces) of spray in the same amount of time it took to run the 163 feet. Be sure tractor is at the same throttle setting. You are now applying 20 gallons per acre on a 20 inch nozzle boom spacing.
   D. For each inch of nozzle spacing on boom, increase time by 5% or reduce the volume by 5%.

---

*If you have calibrated your rig and it is putting out 37 gallons/acre, the required dosage is 4 pounds actual/acre. Therefore, for every 37 gallons of carrier (water, oil, etc.) in the spray tank you add 4 pounds of active material.

USEFUL FORMULAE

1. To determine the amount of active ingredient needed to mix in the spray tank.
   No. of gallons or pounds =

   $$\frac{\text{No. of acres to be sprayed x pounds active ingredient required per A}}{\text{pounds active ingredient per gallon or per pound}}$$

2. To determine the amount of pesticide needed to mix a spray containing a certain percentage of the active ingredient.
   No. of gallons or pounds =

   $$\frac{\text{gallons of spray desired x \% active ingredient wanted x 8.345}}{\text{pounds active ingredient per gallon or pound x 100}}$$

3. To determine the percent active ingredient in a spray mixture.
   Percent =

   $$\frac{\substack{\text{pounds or gallons of concentrate used (not just active ingredient)} \\ \text{x \% active ingredient in the concentrate}}}{\text{gallons of spray x 8.345 (weight of water/gallon)}}$$

4. To determine the amount of pesticide needed to mix a dust with a given percent active ingredient.
   pounds material =

   $$\frac{\text{\% active ingredient wanted x pounds of mixed dust wanted}}{\text{\% active ingredient in pesticide used}}$$

5. To determine the size of pump needed to apply a given number of gallons/acre.
   pump capacity =

   $$\frac{\text{gallons/acre desired x boom width (feet) x mph}}{495}$$

6. To determine the nozzle capacity in gallons per minute at a given rate/acre and miles/hour.
   Nozzle capacity =

   $$\frac{\text{gallons/acre x nozzle spacing (inches) x mph}}{5940}$$

7. To determine the acres per hour sprayed.
   Acres per hour =

   $$\frac{\text{swath width (inches) x mph}}{100}$$

8. To determine the rate of speed in miles per hour.
   1. Measure off a distance of 300 to 500 feet.
   2. Measure in seconds the time it takes the tractor to go the marked off distance.
   3. Multiply .682 times the distance traveled in feet and divide product by the number of seconds.

   $$\text{MPH} = \frac{.682 \text{ x distance}}{\text{seconds}}$$

9. To determine the nozzle flow rate.
    Time the seconds necessary to fill a pint jar from a nozzle.
    Divide the number of seconds into 7.5.
    gallons/minute/nozzle = $\dfrac{7.5}{seconds}$

10. To determine the gallons per minute per boom.
    Figure out the gallons/minute/nozzle and multiply by the number of nozzles.

11. To determine the gallons per acre delivered.
    $$\frac{5940 \times \text{gallons/minute/nozzle}}{\text{nozzle spacing (inches)} \times \text{mph}} = \text{gpa}$$

12. To determine the acreage sprayed per hour.
    acres sprayed/hour = $\dfrac{\text{boom width (feet)} \times \text{mph}}{12}$
    This allows 30% of time for filling, turning, etc.

13. Sprayer Tank Capacity
    Calculate as follows:
    1. Cylindrical Tanks:
        Multiply the length in inches times the square of the diameter in inches and multiply the product by .0034.
        length x diameter squared x .0034 = number of gallons.
    2. Elliptical Tanks:
        Multiply the length in inches times the short diameter in inches times the long diameter in inches times .0034.
        length x short diameter x long diameter x .0034 = number of gallons.
    3. Rectangular Tanks:
        Multiply the length times the width times the depth in inches and multiply the product by .004329.
        length x width x depth x .004329 = number gallons.

14. To determine the acres in a given area.
    Multiply the length in feet times the width in feet times 23. Move the decimal point 6 places to the left to give the actual acres.

CONVERSION TABLES (U.S.)

Linear Measure —
    1 foot = 12 inches
    1 yard = 3 feet
    1 rod = 5.5 yards = 16.5 feet
    1 mile = 320 rods = 1760 yards = 5280 feet

Square Measure —
    1 square foot (sq. ft.) = 144 square inches (sq. in.)
    1 square yard (sq. yd.) = 9 sq. ft.
    1 square rod (sq. rd.) = 272.25 sq. ft. = 30.25 sq. yd.
    1 acre = 43560 sq. ft. = 4840 sq. yds. = 160 sq. rds.
    1 square mile = 640 acres

Cubic Measure —
    1 cubic foot (cu. ft.) = 1728 cubic inches (cu. in.) = 29.922 liquid
        quarts = 7.48 gallons
    1 cubic yard = 27 cubic feet

Liquid Capacity Measure —
    1 tablespoon = 3 teaspoons
    1 fluid ounce = 2 tablespoons
    1 cup = 8 fluid ounces
    1 pint = 2 cups = 16 fluid ounces
    1 quart = 2 pints = 32 fluid ounces
    1 gallon = 4 quarts = 8 pints = 128 fluid ounces

Weight Measure —
    1 pound (lb.) = 16 ounces (oz.)
    1 hundred weight (cwt.) = 100 pounds
    1 ton = 20 cwt. = 2000 pounds

Rates of Application —
    1 ounce/sq. ft. = 2722.5 lbs./acre
    1 ounce/sq. yd. = 302.5 lbs./acre
    1 ounce/100 sq. ft. = 27.2 lbs./acre
    1 pound/100 sq. ft. = 435.6 lbs./acre
    1 pound/1000 sq. ft. = 43.6 lbs./acre
    1 gallon/acre = 3 ounces/1000 sq. ft.
    5 gallons/acre = 1 pint/1000 sq. ft.
    100 gallons/acre = 2.5 gallons/1000 sq. ft. = 1 quart/100 sq. ft.
    100 lbs./acre = 2.5 lbs./1000 sq. ft.

Important Facts —

Volume of sphere = diameter^3 x .5236
Diameter = circumference x .31831
Area of circle = diameter^2 x .7854
Area of ellipse = product of both diameters x .7854
Volume of cone = area of base x 1/3 height
1 cubic foot water = 7.5 gallons = 62.5 pounds
Pressure in psi = height (ft.) x .434
1 acre = 209 feet square
ppm = % x 10,000
% = ppm ÷ 10,000
1% by volume = 10,000 ppm

TABLE OF CONVERSION FACTORS

To Convert From	To	Multiply By
Cubic feet	gallons	7.48
Cubic feet	liters	28.3
Gallons	milliliters	3785
Grams	pounds	.0022
Grams/liter	parts/million	1000
Grams/liter	pounds/gallon	.00834
Liters	cubic feet	.0353
Milligrams/liter	parts/million	1
Milliliters/gallons	gallons	.0026
Ounces	grams	28.35
Parts/million	grams/liter	.001
Parts/million	pounds/million gallons	8.34
Pounds	grams	453.59
Pounds/gallon	grams/liter	111.83

1 gram = .035 ounce
1 kilogram = 2.2 lbs.
1 quintal = 100 kg. = 221 lbs.
1 metric ton = 100 kg. = 2,205 lbs.
1 hectare = 2.5 acres
1 meter = 39.4 inches
1 kilometer = .6 mile

CONVERSION TABLE

1 kilogram (kg) = 1000 grams (g) = 2.2 pounds
1 gram (g) = 1000 milligrams (mg) = .035 ounce
1 liter = 1000 milliliters (ml) or cubic centimeters (cc) = 1.058 quarts
1 milliliter or cubic centimeter = .034 fluid ounce
1 milliliter or cubic centimeter of water weighs 1 gram
1 liter of water weighs 1 kilogram
1 pound = 453.6 grams
1 ounce = 28.35 grams
1 pint of water weighs approximately 1 pound
1 gallon of water weighs approximately 8.34 pounds

1 part per million (ppm) = 1milligram/liter
$\qquad\qquad\qquad\qquad\quad$ = 1 milligram/kilogram
$\qquad\qquad\qquad\qquad\quad$ = .0001 percent
$\qquad\qquad\qquad\qquad\quad$ = .013 ounces in 100 gallons of water

1 percent = 10.000 ppm
$\qquad\quad$ = 10 grams per liter
$\qquad\quad$ = 10 grams per kilogram
$\qquad\quad$ = 1.33 ounces by weight per gallon of water
$\qquad\quad$ = 8.34 pounds/100 gallons of water

.1 percent = 1000 ppm = 1000 milligrams/liter
.01 percent = 100 ppm = 100 milligrams/liter
.001 percent = 10 ppm = 10 milligrams/liter
.0001 percent = 1 ppm = 1 milligram/liter

CHEMICAL ELEMENTS

Name	Symbol	Atomic Weight	Valance
Aluminum	Al	26.97	3
Antomony	Sb	121.76	3, 5
Arsenic	As	74.91	3, 5
Barium	Ba	137.36	2
Bismuth	Bi	209.00	3, 5
Boron	B	10.82	3, 0
Bromine	Br	79.916	1, 3, 5, 7
Cadmium	Cd	112.41	2
Calcium	Ca	40.08	2
Carbon	C	12.01	2, 4
Chlorine	Cl	35.457	1, 3, 5, 7
Cobalt	Co	58.94	2, 3
Copper	Cu	63.57	1, 2
Fluorine	F	19.00	1
Hydrogen	H	1.008	1
Iodine	I	126.92	1, 3, 5, 7
Iron	Fe	55.85	2, 3
Lead	Pb	207.21	2, 4
Magnesium	Mg	24.32	2
Mercury	Hg	200.61	1, 2
Molybdenum	Mo	95.95	3, 4, 6
Nickel	Ni	58.69	2, 3
Nitrogen	N	14.008	3, 5
Oxygen	O	16.00	2
Phosphorus	P	30.98	3, 5
Potassium	K	39.096	1
Selenium	Se	78.96	3
Silicon	Si	28.06	4
Silver	Ag	107.88	1
Sodium	Na	22.997	1
Sulfur	S	32.06	2, 4, 6
Thallium	Tl	204.29	1, 3
Tin	Sn	118.70	2, 4
Titanium	Ti	47.90	3, 4
Uranium	U	238.17	4, 6
Zinc	Zn	65.38	2

WIDTH OF AREA COVERED TO ACRES PER MILE TRAVELED

Width of Strip (feet)	Acres/mile
6	.72
10	1.21
12	1.45
12	1.45
16	1.93
18	2.18
20	2.42
25	3.02
30	3.63
50	6.04
75	9.06
100	12.1
150	18.14
200	24.2
300	36.3

TEMPERATURE CONVERSION TABLE RELATIONSHIP OF CENTIGRADE AND FAHRENHEIT SCALES

°C	°F	°C	°F
-40	-40	20	68
-35	-31	25	77
-30	-22	30	86
-25	-13	35	95
-20	-4	40	104
-15	5	45	113
-10	14	50	122
-5	23	55	131
0	32	60	140
5	41	80	176
10	50	100	212
15	59		

PROPORTIONATE AMOUNTS OF DRY MATERIALS

Water	Quantity of Material				
100 gallons	1 lb.	2 lbs.	3 lbs.	4 lbs.	5 lbs.
50 gallons	8 oz.	1 lb.	24 oz.	2 lbs.	2 1/2 lbs.
5 gallons	3 tbs.	1 1/2 oz.	2 1/2 oz.	3 1/4 oz.	4 oz.
1 gallon	2 tsp.	3 tsp.	1 1/2 tbs.	2 tbs.	3 tbs.

PROPORTIONATE AMOUNTS OF LIQUID MATERIALS

Water	Quantity of Material		
100 gallons	1 qt.	1 pt.	1 cup
50 gallons	1 pt.	1 cup	1/2 cup
5 gallons	3 tbs.	5 tsp.	2 1/2 tsp.
1 gallon	2 tsp.	1 tsp.	1/2 tsp.

MILES PER HOUR CONVERTED TO FEET PER MINUTE

MPH	fpm
1	88
2	176
3	264
4	352

EMULSIFIABLE CONCENTRATE PERCENT RATINGS IN POUNDS ACTUAL PER GALLON

%EC	lbs./Gallon
10-12	1
15-20	1.5
25	2
40-50	4
60-65	6
70-75	8
80-100	10

CONVERSION TABLE FOR LIQUID FORMULATIONS*

Concentration of Active Ingredient in Formulations, lbs./gal.

Rate Desired Lbs./A	1	2	2.5	3	4	5	6
	(ml of formulation per 100 square feet)						
1	8.67	4.33	3.47	2.89	2.17	1.73	1.44
2	17.3	8.67	6.93	5.78	4.33	3.47	2.89
3	26.0	13.0	10.4	8.67	6.50	5.20	4.33
4	34.8	17.4	13.9	11.6	8.69	6.95	5.79
5	43.4	21.7	17.4	14.5	10.0	8.68	7.24
6	52.1	26.0	20.8	17.4	13.0	10.4	8.68
7	60.8	30.4	24.3	20.3	15.2	12.2	10.1
8	69.4	34.7	27.8	23.1	17.4	13.9	11.6
9	78.1	39.0	31.2	26.0	19.5	15.6	13.0
10	86.7	43.3	34.7	28.9	21.7	17.3	14.4

*Example: To put out a 100 sq. ft. plot at the rate of 5 lbs./A active ingredient using a formulation containing 4 lbs./gal. active ingredients, use 10.7 ml. of the 4 lbs./gal. formula and distribute evenly.

CONVERSION TABLE FOR DRY FORMULATIONS

Concentration of Active Ingredient in Formulation

Rate Desired Lbs./A	100%	90%	80%	75%	70%	60%	50%	40%	30%	25%	20%	10%	5%
	(Grams of formulation per 100 square feet)												
1	1.04	1.16	1.30	1.39	1.49	1.74	2.08	2.60	3.47	4.17	5.21	10.4	20.8
2	2.08	2.31	2.60	2.78	2.98	3.47	4.17	5.21	6.94	8.33	10.4	20.8	41.7
3	3.12	3.47	3.90	4.17	4.46	5.20	6.25	7.81	10.4	12.5	15.6	31.2	62.5
4	4.17	4.63	5.21	5.55	5.95	6.94	8.33*	10.4	13.9	16.7	20.8	41.7	83.3
5	5.21	5.79	6.51	6.94	7.44	8.68	10.4	13.0	17.4	20.8	26.0	52.1	104
6	6.25	6.94	7.81	8.33	8.93	10.4	12.5	15.6	20.8	25.0	31.2	62.5	125
7	7.29	8.10	9.11	9.72	10.4	12.1	14.6	18.2	24.3	29.2	36.4	72.9	146
8	8.33	9.26	10.4	11.1	11.9	13.9	16.7	20.8	27.8	33.3	41.7	83.3	167
9	9.37	10.4	11.7	12.5	13.4	15.6	18.7	23.4	31.2	37.5	46.9	93.7	187
10	10.4	11.6	13.0	13.9	14.9	17.4	20.8	26.0	34.7	41.7	52.1	104	208

*Example: To put out a 100 sq. ft. plot at the rate of 4 lbs./A active ingredient using a formulation containing 50% active ingredient, use 8.33 grams of the 50% formulation and distribute evenly over the 100 sq. ft.

CONVERSION TABLE FOR GRANULR FORMULATIONS

Concentration of Active Ingredient in Formulation

Rate Desired Lbs./A	1%	2%	3%	4%	5%	7.5%	10%	15%	20%
1	104.0	52.0	34.66	26.0	20.8	13.86	10.4	6.94	5.2
2	208.0	104.0	69.3	52.0	41.7	27.7	20.8	13.9	10.4
3	312.0	156.0	103.9	78.0	62.5	41.6	31.2	20.8	15.6
4	416.0	208.0	138.6	104.0	83.3	55.4	41.7	27.8	20.8
5	520.0	260.0	173.3	130.0	104.0*	69.3	52.1	34.7	26.0
6	624.0	312.0	207.9	156.0	125.0	83.2	62.5	41.6	31.2
7	728.0	364.0	242.6	182.0	146.0	97.0	72.9	45.6	36.4
8	832.0	416.0	277.3	208.0	167.0	110.9	83.3	55.5	41.7
9	936.0	468.0	311.9	234.0	187.0	124.7	93.7	62.5	46.9
10	1040.0	520.0	346.6	260.0	208.0	168.6	104.0	69.4	52.1
15	1560.0	780.0	519.2	390.0	312.0	207.9	156.0	104.1	78.0
20	2080.0	1040.0	693.2	520.0	416.0	277.2	208.0	138.8	104.0
25	2600.0	1300.0	866.5	650.0	520.0	346.5	260.0	173.5	130.0
30	3120.0	1560.0	1039.8	780.0	624.0	415.8	312.0	208.2	156.0

*Example: To put out a 100 sq. ft. plot at the rate of 5 lbs./A active ingredient using a formulation containing 5% active material, use 104.0 grams of the 5% formulation and distribute it evenly over the 100 sq. ft.

GRAMS/GALLON TABLE

Gallons PPM	5	10	15	20	25	50	75	100	150	200	300	400
5	0.1	0.2	0.3	0.4	0.5	1.0	1.4	1.9	2.8	3.8	5.7	7.6
10	0.2	0.4	0.6	0.8	1.0	1.9	2.8	3.8	5.7	7.6	11.0	15.0
15	0.3	0.6	0.9	1.1	1.4	2.8	4.3	5.7	8.5	11.0	17.0	23.0
20	0.4	0.8	1.1	1.5	1.9	3.8	5.7	7.6	11.0	15.0	23.0	30.0
25	0.5	0.9	1.4	1.9	2.4	4.7	7.1	9.5	14.0	19.0	28.0	38.0
50	0.9	1.9	2.8	3.8	4.7	9.5	14.0	19.0	28.0	38.0	57.0	76.0
75	1.4	2.8	4.3	5.7	7.1	14.0	21.0	28.0	43.0	57.0	85.0	114.0
100	1.9	3.8	5.7	7.6	9.5	19.0	28.0	38.0	57.0	76.0	114.0	151.0
125	2.4	4.7	7.1	9.5	12.0	24.0	36.0	47.0	71.0	95.0	142.0	189.0
150	2.8	5.7	8.5	11.0	14.0	28.0	43.0	57.0	85.0	114.0	170.0	227.0
175	3.3	6.6	9.9	13.0	17.0	33.0	50.0	66.0	99.0	133.0	199.0	265.0
200	3.8	7.6	11.0	15.0	19.0	38.0	57.0	76.0	114.0	151.0	227.0	303.0
250	4.7	9.5	14.0	19.0	24.0	47.0	71.0	95.0	142.0	189.0	284.0	379.0
300	5.7	11.0	17.0	23.0	28.0	57.0	85.0	114.0	170.0	227.0	341.0	454.0
400	7.6	15.0	23.0	30.0	38.0	76.0	114.0	151.0	227.0	303.0	454.0	606.0

DETERMINE THE NUMBER OF ROWS TO THE ACRE

Rows/Acre	Length of Rows				
	32"	36"	38"	40"	60"
1	16335	14520	13756	13068	8712
2	8168	7260	6878	6534	4356
3	5445	4840	4585	4356	2904
4	4084	3630	3439	3267	2178
5	3267	2904	2751	2614	1742
6	2723	2420	2293	2178	1452
7	2334	2074	1965	1867	1245
8	2042	1815	1719	1634	1089
9	1815	1613	1528	1452	968
10	1634	1452	1376	1307	871
11	1485	1320	1251	1188	792
12	1361	1210	1156	1089	726
13	1257	1117	1058*	1005	670
14	1167	1037	982	933	622
15	1089	968	917	871	581
16	1021	908	760	817	545
17	961	854	809	769	512
18	908	807	764	726	484
19	860	764	724	688	459
20	817	726	688	653	436
21	778	691	655	622	415
22	743	660	625	594	396
23	710	631	598	568	379
24	681	605	573	544	363
25	653	581	550	523	348
26	628	558	529	503	335
27	605	538	509	484	323
28	583	519	491	467	311
29	563	501	474	450	300
30	545	484	459	436	290

*Example: A grower's field is 1058 feet long furrowed out on 38-inch centers. Therefore, every 13 rows across the field represents an acre.

QUICK CONVERSIONS

TEMPERATURE			LENGTH			VOLUME	
°C	°F		cm	inch		liters	quarts
100	212		2.5	1		1	1.1
90	194		5	2		2	2.1
80	176		10	4		3	3.2
70	158		20	8		4	4.2
60	140		30	12		5	5.3
50	122		40	16		6	6.3
40	104		50	20		7	7.4
35	95		60	24		8	8.5
30	86		70	28		9	9.5
25	77		80	32			
20	68		90	36			
15	59		100	39			
10	50		200	79			
5	41			feet			
0	32		300	10			
-5	23		400	13			
-10	14		500	16			
-15	5		1,000	33			
-20	-4						
-25	-13						
-30	-22						
-40	-40						

QUICK CONVERSIONS

kg./ha.		lb./A
1	....	0.9
2	....	1.8
3	....	2.7
4	....	3.6
5	....	4.5
10	....	9
20	....	18
20	....	27
40	....	36
50	....	45
60	....	54
70	....	62
80	....	71
90	....	80
100	....	89
200	....	180
300	....	270
400	....	360
500	....	450
600	....	540
700	....	620
800	....	710
900	....	800
1000	....	890
2000	....	1800

		ton/A
3000	....	1¼
4000	....	1¾
5000	....	2¼
6000	....	2¾
7000	....	3
8000	....	3½
9000	....	4
10000	....	4½
11000	....	5
12000	....	5½
13000	....	5¾
14000	....	6¼
15000	....	6¾
16000	....	7
17000	....	7½
18000	....	8
19000	....	8½
20000	....	9

USEFUL MEASUREMENTS

LENGTH
1 mile = 80 chains = 8 furlongs = 1760 yards = 5280 feet = 1.6 kilometers
1 chain = 22 yards = 4 rods, poles or perches = 100 links

WEIGHT
1 long ton = 20 cwt. = 2240 pounds
1 pound = 16 ounces = 454 grams = 0.454 kilograms
1 short ton = 2000 pounds
1 metric ton = 2204 pounds = 1000 kilograms

AREA
1 acre = 10 sq. chains = 4840 sq. yards = 43560 sq. ft. = 0.405 hectares
1 sq. mile = 640 acres = 2.59 kilometers
1 hectare = 2.471 acres

VOLUME
1 gal. = 4 quarts = 8 pints = 128 fluid ozs. = 3.785 liters
1 fluid oz. = 2 tablespoons = 4 dessertspoons = 8 teaspoons = 28.4 c.c.'s

CAPACITIES
Cylinder — diameter $^2/$ x depth x 0.785 = cubic feet
Rectangle — breadth x depth x length = cubic feet
Cubic feet x 6.25 = gallons

QUICK CONVERSIONS
1 pint/acre
1 gal./acre
1 lb./acre
1 cwt./acre   1 acre = 1 fluid oz./242 sq. yards
1 m.p.h.   = 1 pint/605 sq. yards
3 m.p.h.   = 1 oz./300 sq. yards
1 liter/hectare   = 0.37 oz./sq. yard
1 kilogram/hectare   = 88 ft./minute
1 c.c./100 liters   = 1 chain/15 sec.
125 c.c./100 liters   = 0.089 gal./acre
1 gm./100 liters   = 0.892 lb./acre
      = 0.16 fl. oz./100 gallons
      = 1 pint/100 gallons
A strip 3 ft. wide x 220 chains   = 0.16 oz./100 gallons
A strip 4 ft. wide x 165 chains
A strip 5 ft. wide x 132 chains

CONVERSION FACTORS USED IN CALCULATION

Convert	To	By
gram (gm.)	kilogram (kg.)	move decimal 3 places to left

Example: 2000 gm. = 2.0 kg.

gram (gm.)	milligram (mg.)	move decimal 3 places to right

Example: 2.0 gm. = 2000 mg.

gram (gm.)	pound (lb.)	divide by 454

Example: 658 gm./÷454 = 1.45 lb.

gram/pound	percent (%)	divide by 4.54

Example: 90 gm./lb. ÷ 4.54 = 19.8%

gram/ton	percent	multiply by 11, move decimal 5 places to left

Example: 45 gm./ton x 11 = 495 = .00495%

kilogram (kg.)	gram	move decimal 3 places to right

Example: 5.5 kg. = 5500 gm.

milligram (mg.)	gram	move decimal 3 places to left

Example: 95 mg. = 0.095 gm.

percent	gram/pound	multiply by 4.54

Example: 25 x 4.54 = 113.5 gm./lb.

percent	parts/million (ppm)	move decimal 4 places to right

Example: .025% = 250 ppm.

percent	gram/ton	divide by 11, move decimal 5 places to right

Example: .011 ÷ 11 = .001 = 100 gm./ton

pound	gram	multiply by 454

Example: 0.5 lb. x 454 = 227 gm.

ppm	percent	move decimal 4 places to left

Example: 100 ppm = 0.01%

290

STANDARD MEASUREMENTS

MEASURE OF LENGTH (Linear Measure)

4 inches	=	1 hand
9 inches	=	1 span
12 inches	=	1 foot
3 feet	=	1 yard
6 feet	=	1 fathom
5½ yards - 16½ feet	=	1 rod
40 poles	=	1 furlong
8 furlongs	=	1 mile
5,280 feet = 1,760 yards	=	320 rods = 1 mile
3 miles	=	1 league

MEASURE OF SURFACE (area)

144 square inches	=	1 square foot
9 square feet	=	1 square yard
30¼ square yards	=	1 square rod
160 square rods	=	1 acre
43,560 square feet	=	1 acre
640 square acres	=	1 square mile
36 square miles	=	1 township

SURVEYOR'S MEASURE

7.92 inches	=	1 link
25 links	=	1 rod
4 rods	=	1 chain
10 square chains	=	160 square rods = 1 acre
640 acres	=	1 square mile
80 chains	=	1 mile
1 Gunter's chain	=	66 feet

METRIC LENGTH

1 inch	=	2.54 centimeters
1 foot	=	.305 meter
1 yard	=	.914 meter
1 mile	=	1.609 kilometers
1 fathom	=	6 feet
1 knot	=	6,086 feet
3 knots	=	1 league
1 centimeter	=	.394 inch
1 meter	=	3.281 feet
1 meter	=	1.094 yards
1 kilometer	=	.621 mile

TROY WEIGHT

24 grains	=	1 pennyweight
20 pennyweight	=	1 ounce
12 ounces	=	1 pound

LIQUID MEASURE

2 cups	=	1 pint
4 gills	=	1 pint
16 fluid ounces	=	1 pint
2 pints	=	1 quart
4 quarts	=	1 gallon
31½ gallons	=	1 barrel
2 barrels	=	1 hogshead
1 gallon	=	231 cubic inches
1 cubic foot	=	7.48 gallons
1 teaspoon	=	.17 fluid ounces (1/6 oz.)
3 teaspoons (level)	=	1 tablespoon (1/2 oz.)
2 tablespoons	=	1 fluid ounce
1 cup (liquid)	=	16 tablespoons (8 oz.)
1 teaspoon	=	5 to 6 cubic centimeters
1 tablespoon	=	15 to 16 cubic centimeters
1 fluid ounce	=	29.57 cubic centimeters

CUBIC MEASURE (Volume)

1,728 cubic inches	=	1 cubic foot
27 cubic feet	=	1 cubic yard
2,150.42 cubic inches	=	1 standard bushel
231 cubic inches	=	1 standard gallon (liquid)
1 cubic foot	=	4/5 of a bushel
128 cubic feet	=	1 cord (wood)
7.48 gallons	=	1 cubic foot
1 bushel	=	1.25 cubic feet

DRY MEASURE

2 pints	=	1 quart
8 quarts	=	1 peck
4 pecks	=	1 bushel
36 bushels	=	1 chaldron

APOTHECARIES' WEIGHT

20 grains	=	1 scruple
3 scruples	=	1 dram
8 drams	=	1 ounce
12 ounces	=	1 pound
27-11/32 grains	=	1 dram
16 drams	=	1 ounce
16 ounces	=	1 pound
2,000 pounds	=	1 ton (short)
2,240 pounds	=	1 ton (long)

CONVERSION FACTORS

Degree C = 5/9 (Degree F − 32).

Degree F = 9/5 (Degree C + 32).

Degrees Absolute (Kelvin) = Degrees centigrade + 273.16.

Degrees absolute (Rankine) = Degrees fahrenheit + 459.69.

Multiply	By	To Obtain
Diameter circle	3.1416	Circumference circle
Diameter circle	0.8862	Side of equal square
Diameter circle squared	0.7854	Area of circle
Diameter sphere squared	3.1416	Area of sphere
Diameter sphere cubed	0.5236	Volume of sphere
U.S. Gallons	0.8327	Imperial gallons (British)
U.S. Gallons	0.1337	Cubic feet
U.S. Gallons	8.330	Pounds of water (20° C)
Cubic feet	62.427	Pounds of water (4° C)
Feet of water (4° C)	0.4335	Pounds per square inch
Inch of mercury (0° C)	0.4912	Pounds per square inch
Knots	1.1516	Miles per hour

Figuring Grain Storage Capacity

1 bu. ear corn = 70 lbs. 2.5 cu. ft. (15.5% moisture)

1 bu. shelled corn = 56 lbs. 1.25 cu. ft. (15.5% moisture)

1 cu. ft. = 1/2.50 = .4 bu. of ear corn

1 cu. ft. = 1/1.25 = .8 bu. of shelled corn; Bu. x 1.25 ft.³, ft.³ x .8 bu.

Ft.³ = Bu. x 1.25

Bu. = Ft.³ x .8

Rectangular or square cribs or bins

   cu. ft. = width x height x length (W x H x L)

Round cribs, bins or silos (= 3.1416)

   Volume = R²H = D²H/r

   cu. ft. = x radius x radius x height = (R x R x H)

   or x diameter x diameter x height = (D x D x H)

   $$\frac{\text{diameter} \times \text{diameter} \times \text{height}}{4}$$

   or $\dfrac{3.1416 \times D \times D \times H}{4}$ = .785 x D x D x H

Examples

1. Crib — ear corn — 6' wide by 12' high by 40' long
   a. 6 x 12 x 40 = 2880 cu. ft. x 4 bu./cu. ft. = 1152 bu.
   b. 6 x 12 x 1 = 72 cu. ft. x .4 28.8 bu./ft. of length x 40' = 1152 bu.

2. Round crib — ear corn — 14' diameter by 14' high
   a. .785 x 14' x 14' x 14' x .8 = 1722 bushel
   b. .785 x 14' x 1 x .4 6.15 bu./ft. x 14 = 861 bushel

3. Round Bin or Silo — shell corn — 14' diameter by 14' high
   a. .785 x 14' x 14' x 14' x .8 = 1722 bushel
   b. .785 x 14' x 14' 14' x 1 x .8 - 123 bu./ft. x 14' = 1722 bushel

Metric Weight

1 grain	=	.065 gram
1 apothecaries' scruple	=	1.296 grams
1 avoirdupois ounce	=	28.350 grams
1 troy ounce	=	31.103 grams
1 avoirdupois pound	=	.454 kilogram
1 troy pound	=	.373 kilogram
1 gram	=	15.432 grains
1 gram	=	.772 apothecaries' scruple
1 gram	=	.035 avoirdupois ounce
1 gram	=	.032 troy ounce
1 kilogram	=	2.205 avoirdupois pounds
1 kilogram	=	2.679 troy pounds

Capacity

1 U.S. fluid ounce	=	29.573 ml
1 U.S. fluid quart	=	.946 liter
1 U.S. fluid ounce	=	29,573 milliliters
1 U.S. liquid quart	=	.964 liter
1 U.S. dry quart	=	1.101 liters
1 U.S. gallon	=	3.785 liters
1 U.S. bushel	=	.3524 hectoliters
1 cubic inch	=	16.4 cubic centimeters
1 liter	=	1,000 milliliters or 1,000 cubic centimeters
1 cubic foot water	=	7.48 gallons or 62-1/2 pounds
231 cubic inches	=	1 gallon
1 milliliter	=	.034 U.S. fluid ounce
1 liter	=	1.057 U.S. liquid quarts
1 liter	=	.908 U.S. dry quart
1 liter	=	.264 U.S. gallon
1 hectoliter	=	2.838 U.S. bushels
1 cubic centimeter	=	.061 cubic inch

Miscellaneous Equivalents

9 in. equals 1 span

6 ft. equals 1 fathom

6,080 ft. equals 1 nautical mile

1 board ft. equals 144 cu. in.

1 cylindrical ft. contains 4-7/8 gals.

1 cu. ft. equals .8 bushel

12 dozen (doz.) equals 1 gross (gr.)

1 gal. water weighs about 8-1/3 lbs.

1 gal. milk weighs about 8.6 lbs.

1 gal. cream weighs about 8.4 lbs.

46-1/2 qts. of milk weighs 100 lbs.

1 cu. ft. water weighs 62-1/2 lbs., contains 7-1/2 gals.

1 gal. kerosene weighs about 6-1/2 lbs.

1 bbl. cement contains 3.8 cu. ft.

1 bbl. oil contains 42 gals.

1 standard bale cotton weighs 480 lbs.

1 keg of nails weighs 100 lbs.

4 in. equals 1 hand in measuring horses

ADDRESS OF
BASIC MANUFACTURERS

Abbott Laboratories
Abbott Park-1400 Sheridan Road
N. Chicago, IL 60064

Agbiochem Inc.
3 Fleetwood Court
Orinda CA 94563

AgriDyne Technologies
417 Wakara Way
Salt Lake City, Utah 84108

Agrisence
4230 W. Swift #106
Fresno, CA 93722

Agro-Kanesho Co. Ltd.
1-1 Marunonchi 3-chome
Chiyodi-ku
Tokyo 100 Japan

Agrolinz
Agrarchemihalein GmbH
St. Peter Strasse 25
POB 21
A-4021 Linz, Austria

Agtrol Chemical Products
7322 Southwest Freeway,
Ste 1400
Houston, TX 77074

Agway Inc.
P.O. Box 4933
Syracuse, NY 13221-4933

Albaugh Chemical Corp.
728 SE Creek View Dr.
Ankeny, IA 50021

American Cyanamid Co.
One Cyanamid Plaza
Wayne, NJ 07470

Amvac Chemical Corp.
4100 E. Washington Blvd.
Los Angeles, CA 90023

Applied Biochemists
6120 W. Douglas Ave.
Milwaukee WI 53218

Asahi Chemical Industry Co.
500 Oaza-Takayasu
Ikarugacho, Ikomagun
Nara 636-01 Japan

Atochem North America
Ag Chemical Division
Three Parkway Rm. 619
Philadelphia PA 19102

Avitrol Corp.
7644 E. 46th St.
Tulsa, OK 74145

Bactec Corp.
9601 Katy Freeway Ste. 350
Houston, TX 77024-1333

BASF
Postfach 220
6703 Limburgerhof
Germany

BASF
Ag Chemical Div.
P.O. Box 13528
Research Triangle Park, NC 27709

Bayer AG
5090 Leverkusen
Bayerwerk,
Germany

Bell Labs
3699 Kinsman Boulevard
Madison, WI 53704

Bentech Laboratories
4370 NE Halsey St.
Portland, OR 97213

Bernado Chemicals
3994 Swinnea Rd.
Memphis, TN 38118

Biocontrol Ltd.
719 Second St., Suite 12
Davis, CA 95616

Biologic Inc.
11 Lake Ave. Extension
Danbury, Conn. 06811

Biosys
1057 E. Meadow Circle
Palo Alto, CA 94303

Boehringer Ingelheim Animal Health
2621 N. Belt Highway
St. Joseph, MO 64506

Boliden Intertrade Inc.
3400 Peachtree Rd. NE
Suite 401
Atlanta, GA 30326

Brandt Consolidated
P.O. Box 277
Pleasant Plains, IL 62677

Brogdex Co.
1441 W. 2nd St.
Pomona, CA 91766

Buckman Labs
1256 N. McLean Blvd.
Memphis, TN 38108

C.P. Chemicals, Inc.
1 Parker Place
Ft. Lee, NJ 07224

Calliope S.A.
7 Rue du Chapeau Rouge
34500 Beziers, France

Cedar Chemical Co.
5100 Poplar Ave.
24th Floor
Memphis TN 38137

Cenex/Land o Lakes
P.O. Box 64089
St. Paul, MN 55164-0089

CFPI
P.O. Bix 75
92233 Gennevilliers France

Chemolimpex
Plant Protection Dept.
P.O. Box 121
H-1805 Budapest,
Hungary

Cheminova Agro
P.O. Box 9
DK-7620 Lemvig
Denmark

Cheminova Inc.
Oak Park Hill
1700 Rte 23 Ste 210
Wayne, NJ 07470

Chugai Pharmaceutical Co. Ltd.
Meiho Bldg.
21-1 Nish-Shinjuku 1-Chome
Shinjuku-ku
Tokyo 160 Japan

CIBA-Geigy AG
CH-4002 Basel 7
Switzerland

CIBA-Geigy Corporation
Agricultural Chemicals
P.O. Box 18300
Greensboro, NC 27419

W. A. Cleary Company
Southview Industrial Park
178 Rte 522 Ste A
Dayton, NJ 08810

Concep Membranes Inc.
213 SW Columbia
Bend OR 97703

Coopers Animal Health
2000 S. 11th St.
Kansas City, KS 66103-1438

Cornbelt Chemical Co.
P.O. Box 410
McCook, NE 69001

Crop Genetics Intl.
7170 Standard Dr.
Hanover, MD 21076

CuproQuim Corp.
9601 Katy Freeway Ste. 350
Houston TX 77024-1333

Burlington Bio Medical Corp
222 Sherwood Ave.
Farmingdale, New York 11735-1718

Dainippon Ink & Chemicals
DIC Building
7-20 Nihonbashi 3 Chome
Chuo-ku
Tokyo 103 Japan

Degesch America, Ind.
P.O. Box 116
Weyers Cave, VA 24486

Denka Intl. Inc.
P.O. Box 337
3770 Ah Barneveld
The Netherlands

Detia Degesch
Dr. Werner Freyberg Strasse 1
6947 Laudenbach Bergstrasse
Germany

Dow Elanco
9002 Purdue Rd.
Indianapolis, IN 46268-1189

Dr. R. Maag Ltd.
Chemical Works
CH-8157
Dielsdorf, Switzerland

Drexel Chemical Co.
P.O. Box 9306
Memphis TN 38109-0306

Dunhill Chemical Co.
3026 Muscatel Ave.
Rosemead, CA 91770

DuPont Company
Agricultural Chemical Dept.
Barley Mill Plaza
P.O. Box 80038
Wilmington, DE 19898

Eastman Kodak Co.
Life Sciences Div.
343 State Street
Rochester, NY 14650

J. T. Eaton & Co. Inc.
1393 E. Highland Road
Twinsburg OH 44087

Ecogen
2005 Cabot Blvd., West
Langhorn, PA 19047-1810

Eco Science
1 Innovation Dr.
Worcester, MA 01605

Endura S.P.A.
Viale Pietramellara 5
40121 Bologna Italy

Enichem Agricoltura
Via Medici Del Vascella 40C
21038 Milano Italy

FMC Corporation
Ag Chemical Div.
1735 Market Street
Philadelphia, PA 19103

Fair Products Inc.
P.O. Box 386
Cary, NC 27512-0386

Farmland Industries
P.O. Box 7305
Kansas City, MO 64116-0005

Fermenta Animal Health Co.
10150 N. Executive Hills Blvd.
Kansan City, MO 64153-2315

Fermone Corp. Inc.
2620 37th Dr. N.
Phoenix, AZ 85009

Gowan Co.
P.O. Box 5569
Yuma AZ 85366-5569

Grace-Sierra Horticultural
Products
1001 Yosemite Dr.
Milpitas, CA 95035

Great Lakes Chemical Corp.
P.O. Box 2200
W. Lafayette, IN 47906

Griffin Corporation
P.O. Box 1847
Valdosta, GA 31603

Gustafson, Inc.
1400 Preston Rd., Ste. 400
Plano, TX 75093

Hacco Inc.
P.O. Box 7190
Madison, WI 53707

Helena Chemical Co.
6075 Poplar Ave.
Suite 500
Memphis, TN 38119

Hendrix & Dail Inc.
P.O. Box 648
Greenville, NC 27835-0648

Hercon Environmental Co.
Aberdeen Road
Emigsville, PA 17318

Hess & Clark
7th and Orange Sts.
Ashland, OH 44805

Hodogaya Chemical
1-4-2 Toranomon-1-Chome
Minato-ku
Tokyo 105 Japan

Hoechst-Agrochem. Div.
Postfach 80 03 20
6230 Frankfurt (m) 80
Germany

Hoechst-Roussel Agri.
Vet. Co.
Route 202-206 North
Soverville, NJ 08876-1258

Hokko Chemical Industries
Mitsui Building 2
4-20 Nihonbashi Hongoku-cho 4-chone
Tokyo 103 Japan

ICI Americas, Inc.
Ag Products Division
Wilmington, DE 19897

ICI Ltd.
Plant Protection Div.
Fernhurst, Hasslemere 7
Surrey, England GU2 3JE

Igene BioTechnology Inc.
9110 Red Branch Road
Columbia MD 21045

Ihara Chemical Co.
1-4-26 Ikenohata 1-Chome
Taitoku
Tokyo 110 Japan

Ishihara Sangyo AgroLtd.
10-30 Fujimi 2-Chome
Chiyoda-ku, Tokyo 102
Japan

ISK Biotech
5966 Heisley Rd.
Mentor, OH 44061-8000

Janssen Pharmaceutical
Plant Protection Div.
1125 Trenton-Harbouton Rd.
Titusville, NJ 08560

Janssen Pharmaceutica N. V.
Agricultural Div.
B-2340 Beerse, Belgium

Kaken Pharmaceutical Co., Ltd.
Mitsuihoncho Bldg.
4-10 Nihonbashi honcho 3-chone
Chuo-ku
Tokyo 103 Japan

Kemira Oy
P.O. Box 44
SF-02271 Espoo Finland

Kincaid Enterprises
P.O. Box 549
Nitro, WV 25143

Kumiai Chemical Industries
4-26 Ikenohata 1-Chome
Tokyo 110 Japan

Kureha Chemical Ind. Co.
1-9-11 Nihonbashi,
Horidome-cho, Chuo-ku,
Tokyo 103 Japan

Lebanon Agri Corp.
P.O. Box 180
Lebanon, PA 17042-0180

LESCO Inc
20005 Lake Road
Rocky River, OH 44116

Lipha Tech
34 Rue Saint Romain
69008 Lyon, France

Lipha Tech
3101 W. Custer Ave.
Milwaukee, WI 53209

Loveland Industries
P.O. Box 1289
Greeley, CO 80632

Luxenbourg Industries
P.O. Box 13
Tel Aviv 61000 Israel

Luxembourg-Pamol Inc.
5100 Poplar Ave. Ste. 2746
Memphis Tenn. 38137

Maag Agrochemicals
P.O. Box 6430
Vero Beach, FL 32961-6430

MAGNA-Herbicide Div.
P.O. Box 11192
Bakersfield, CA 93389

A. H. Marks & Co.
Wyke Bradford
West Yord BD12 9EJ England

Maktheshim-Agan
551 Fifth Ave., Ste. 1100
New York, NY 10176

Maktheshim-Agan
P.O. Box 60
84100 Beer-Sheva, Israel

McLaughlin Gromley King
8810-10th Ave., North
Minneapolis, MN 55427

Meiji Seika Company
4-16 Kyobashi 2-Chome
Chuo-ku
Tokyo 104 Japan

Merck-Ag Vet
P.O. Box 2000
Rahway, NJ 07065-6430

E. Merck A. G.
61 D Armstadt
Franfurter Strasse 250
Germany

Micro-Flo Co.
P.O. Box 5948
Lakeland, FL 33807

Miles Inc.
P.O. Box 4913
Kansas City, MO 64120-6013

Miles Animal Health
P.O. Box 390
Shawnee Mission, KS 66201-0390

Miller Chemical & Fert. Corp.
Box 333
Hanover, PA 17331

Minerals Res. & Devel. Corp.
1Woodlawn Green
Suite 232
Charlotte, NC 28217

Mitsubishi Kasei Corp.
Agric. Chemical Div.
Mitsubishi Shozi Bldg.
5-2, Marunouchi 2-Chome
Chiyoda-ku
Tokyo 100 Japan

Mitsubishi Petrochemical Co. Ltd.
2-5-2 Marunouchi, 2-Chome
Chiyoda-ku
Tokyo 100 Japan

Mitsui Agricultural Chemicals
1-2-1 Ohtemachi, Chiyoda-ku,
Tokyo 100 Japan

Mitsui Toatsu Chemicals
Kasumigaschi Building
2-5 Kasamigaseki 3-chome
Chuyoda-hu
Tokyo 100 Japan

Monsanto Chemical Company
800 N. Lindburgh Blvd.
St. Louis, MO 63167

Monterey Chemical Co.
P.O. Box 5317
Fresno, CA 93755

Motomco Ltd.
29 N. Ft. Harrison Ave.
Clearwater, FL 33615

MTM Agrochemicals Ltd.
18 Liverpool Road, Great Sankey
Warrington, Cheshire England

Mycogen Corp.
5451 Oberlin Dr.
San Diego, CA 92121

Nihon Bayer Agrochem
Itopia Nihonbashi Honcho Bldg.
7-1 Nihonbashi Honcho 2-chome
Chuo-hu
Tokyo 103 Japan

Nihon Nohyaku Company, Ltd.
2-5 Nihonbashi 1-Chome
Chuo-ku
Tokyo 103 Japan

Nippon Kayaku Co.-Ag Div.
Tokyo Kaijo Bldg.
2-1, Marunouchi 1-Chome
Chiyoda-ku
Tokyo 100 Japan

Nippon Soda Co., Ltd.
New-Ohtemachi Bldg.
2-1, 2-Chome Ohtemachi
Chiyoda-ku
Tokyo 100 Japan

Nissan Chemical Ind., Ltd.
Kowa-Hitotsubashi Bldg.
7-1, 3-Chome, Kanda
Nishiki-cho Chiyoda-ku
Tokyo 101 Japan

Nor-Am Chemical Co.
3509 Silverside Rd.
Wilmington, DE 19803

Nordox A/S
Ostensjoveien 13
0661 Oslo 6 Norway

Novo Nordisk
Plant Protection Div.
33 Turner Road
Danbury CT 06813-1907

Old Bridge Chemicals
P.O. Box 194
Old Bridge, NJ 08857

Olympic Chemical Co.
P.O. Box K
Mainland, PA 19451

Ore-Calif. Chemicals Inc.
29454 Meadow View Rd.
Junction City, OR 97448

Otsuka Chemical Co.
2-27 Ohtedohir 3-chome
Chuo-ku
Osaka 540 Japan

PBI/Gordon Corp.
P.O. Box 4090
Kansas City, MO 64101

Pace National Corp.
500 7th Ave. South
Kirkland, WA 98033

Pestcon Systems
5511 Capital Center Dr. Ste 302
Raleigh, NC 27606-3365

Phelps Dodge Refining Corp.
P.O. Box 20001
El Paso, TX 79998

Phillips Petroleum
Bartlesville, OK 74004

Philom Bios Inc.
15 Innovation Blvd.
Saskatoon SK S7N2X8 Canada

Plant Health Technologies
P.O. Box 15057
Boise, ID 83715

Platte Chemical Co.
P.O. Box 667
Greeley, CO 80632

Prentiss Drug &
Chemical Co., Inc.
21 Vernon St. CB 2000
Floral Park, NY 11001

Ralston Purina Company
Checkerboard Square
St. Louis, MO 63188

Regal Chemical Co.
P.O. Box 900
Alpharetta, GA 30239

Rentokil Laboratories
Felcourt, East Grinstead
Sussex, England

Rhone Poulenc
P.O. Box 12014
2 TW Alexander Dr.
Research Triangle Park, NC 27709

Rhone Poulenc Agrochemie
14-20 rue Pierre Baizet
69263 Lyon, France BP 9163

Richland Corp.
686 Passaic Ave.
West Caldwell, NJ 07006

.̇ 189
E̤   .er KY 40010

Ringer Corp.
9959 Valley View Rd.
Minneapolis, Minn. 55344

Riverdale Chemical Co.
425 W. 194th St.
Glenwood IL 60425-1584

Rohm & Haas Company
Independence Mall West
Philadelphia, PA 19105

Roussel Environmental Health
400 Sylvan Ave.
Englewood Cliffs, NJ 07632

Roussel UCLAF
163, Ave. Gambetta
75020 Paris, France

Sandoz Agro
1300 E. Touhy Ave.
Des Plaines, IL 60018

Sandoz Agro
Agrochemical Dept.
Basel, Switzerland CH-4002

Sankyo Co., Ltd.
No. 7-12, Ginza, 2-Chome
Chuo-ku
Tokyo 104 Japan

Sapporo Breweries, Ltd.
10-1 Ginza 7-Chome
Chuo-ku,
Tokyo 104 Japan

SARIAF
20124 Milano
Italy

Scentry Inc.
610 Central Ave.
Billings, MT 59109

Schering AG
Postfach 650311
D-1000 Berlin 65,
Germany

The O.M. Scotts & Sons Co.
14111 Scottslawn Rd.
Marysville OH 43041

SDS Biotech KK
Higashi Shinbashi Bldg.
12-7 Higashi Shinbashi 2-Chome
Minato-ku
Tokyo 105 Japan

Shell Forsching
Postfach 100
D-6501
Schwabenheim
Germany

Shell International Ltd.
Agrochemical Div.
Shell Centre
London SE1 7PG
England

Shin-Etsu Chemical Intl.
Asahi Tokai Bldg.
6-1 Ohtemachi 2-chome
Chuydo-ku
Tokyo 100 Japan

Shionogi and Co., Ltd.
1-8 Doshomachi 3-Chome
Osaka, 541 Japan

J.R. Simplot Co.
P.O. Box 198
Lathrop, CA 95330

SKW Trostberg Ag
Postfach 1262
D-8223 Trostberg
Germany

Sostram Corp.
70 Mansell Ct., Ste 230
Roswell, GA 30076

Source Technology Bilologicals
3355 Hiawatha Ave. S.
Minneapolis MN 55406

Southern Agricultural Insecticide
P.O. Box 218
Palmetto, FL 34220

Southern Mill Creek Products
5414 N. 56 St.
Tampa, FL 33610

Sumitomo Corp.
5-33 4-chome
Kitahama Chuo-ku Osaka 541
Japan

Sumitomo Chemical Americas
1330 Dillon Heights Ave.
Baltimore, MD 21258

Summit Chemical Co.
7657 Canton Center Dr.
Baltimore MD 21224

Sureco
P.O. Box 938
Fort Valley GA 31030

Takeda Chemical Industries
12-10 Nihonbashi 2-Chome
Chuo-ku
Tokyo 103 Japan

Terra International Inc.
600 4th St.
Sioux City, IA 51101

Tosoh Corporation
1-7-7 Akasaka
Minato-ku,
Tokyo 107 Japan

Trical
P.O. Box 1327
Hollister, CA 95024-1327

UBE Ind. Ltd.
UBE Bldg.
311 Higashi-shingawa 2-chome
Shinagawa-ku
Tokyo 140 Japan

UCB Chemicals Corp.
5505-A Robin Hood Rd.
Norfolk, VA 23513

Uniroyal Chemical Co.
Crop Protection Division
Middlebury, CT 06749

United Agri Products Inc.
P.O. Box 1286
Greeley, CO 80632

Unocal Chemical Corp.
P.O. Box 60455
Los Angeles, CA 90060

US Borax & Chemical Corp.
3075 Wilshire Blvd.
Los Angeles, CA 90010

Valent Corp.
1333 W. California Blvd.
Walnut Creek, CA 94596-8925

Van Diest Supply
P.O. Box 610
Webster City, IA 50595-0610

Vineland Chemical Company
1611 West Wheat Road
Vineland, NJ 08360

...emie

...nten Strasse 22

L ... J Munchen 22 Germany

Webb Wright Corp.
P.O. Box 1572
Ft. Myers, FL 33902

Westbridge Agricultural Products
2776 Loker Ave. West
Carsbad CA 92008

Western Farm Service
P.O. Box 1168
Fresno, CA 93715

Whitmire Research Labs Inc.
3568 Tree Court
St. Louis, MO 63122-6620

Wilbur Ellis Co.
191 W. Shaw Ave.
Suite 107
Fresno, CA 93704-2876

Zoecon Corporation
12005 Ford Rd.
Suite 800
Dallas TX 75234-7296